Pocket Guide to

Chemical Engineering

Pocket Guide to Chemical Engineering

Carl R. Branan

Gulf Publishing Company
An Imprint of Elsevier
Houston, Texas

Dedication

To my four grandchildren:

Katherine, Alex, Richard, and Matthew.

Pocket Guide to Chemical Engineering

Gulf Publishing Company
An Imprint of Elsevier
Book Division
P. O. Box 2608 □ Houston, Texas 77252-2608

10 9 8 7 6 5 4 3 2

Library of Congress Cataloging-in-Publication Data

Branan, Carl.
Pocket guide to chemical engineering / Carl R. Branan.
p. cm.
Includes bibliographical references and index.
ISBN-13: 978-0-88415-311-5 ISBN-10: 0-88415-311-8 (acid-free paper)
1. Chemical engineering—Handbooks, manuals, etc.
I. Title. II. Title: Chemical engineering.
TP151.B78 1999
660 21—dc21 99-045424

Printed in the United States of America.
Printed on acid-free paper (∞).

Contents

Introduction

This pocket guide contains selected rules of thumb and shortcut design methods meant to travel into the field as well as the office, even when the "office" is a hotel room. It contains updates on certain fast-moving technology and new material not found elsewhere. Miniaturization and easy retrieval of information are stressed. Those on the go can produce reasonable results quickly when using this book as a basic source.

Carl Branan
El Paso, Texas

1
Fluid Flow

GENERAL

Two of the most useful and basic equations are

$$\Delta h = \frac{u^2}{2g} \tag{1-1}$$

$$\Delta P\,(V) + \frac{\Delta u^2}{2g} + \Delta Z + E = 0 \tag{1-2}$$

where Δh = head loss in feet of flowing fluid
- u = velocity in ft/sec
- g = 32.2 ft/sec^2
- P = pressure in lb/ft^2
- V = specific volume in ft^3/lb
- Z = elevation in feet
- E = head loss due to friction in feet of flowing fluid

In Equation 1-1 Δh is called the "velocity head." This expression has a wide range of utility not appreciated by many. It is used "as is" for

1. Sizing the holes in a sparger
2. Calculating leakage through a small hole
3. Sizing a restriction orifice
4. Calculating the flow with a pitot tube

With a coefficient it is used for

1. Orifice calculations
2. Relating fitting losses
3. Relief valve sizing
4. Heat exchanger tube leak calculations

For a sparger consisting of a large pipe having small holes drilled along its length, Equation 1-1 applies directly. This is because the hole diameter and the length of fluid travel passing through the hole are similar dimensions. An orifice, on the other hand, needs a coefficient in Equation 1-1 because hole diameter is a much larger dimension than length of travel (say ⅛ in. for many orifices). Orifices will be discussed under "Metering" later in this chapter.

For compressible fluids one must be careful that when sonic or "choking" velocity is reached, further decreases in downstream pressure do not produce additional flow. This occurs at an upstream to downstream absolute pressure ratio of about 2:1. Critical flow due to sonic velocity has practically no application to liquids. The speed of sound in liquids is very high. See "Sonic Velocity" in this chapter.

Still more mileage can be obtained from $\Delta h = u^2/2g$ when using it with Equation 1-2, which is the famous Bernoulli equation. The terms are

1. The PV change
2. The kinetic energy change or "velocity head"
3. The elevation change
4. The friction loss

These contribute to the flowing head loss in a pipe. However, there are many situations where by chance, or on purpose, $u^2/2g$ head is converted to PV or vice versa.

We purposely change $u^2/2g$ to PV gradually in the following situations:

1. Entering phase separator drums to cut down turbulence and promote separation
2. Entering vacuum condensers to cut down pressure drop

We build up PV and convert it in a controlled manner to $u^2/2g$ in a form of tank blender. These examples are discussed under appropriate sections.

PIPING PRESSURE DROP

A handy relationship for turbulent flow in commercial steel pipes flowing full is:

$$\Delta P_F = W^{1.8}\mu^{0.2}/20{,}000d^{4.8}\rho \tag{1-3}$$

where ΔP_F = frictional pressure loss, psi/100 equivalent ft of pipe
W = flow rate, lb/hr
μ = viscosity, cp
ρ = density, lb/ft^3
d = internal pipe diameter, in.

This relationship holds for a Reynolds number range of 2,100 to 10^6. For smooth tubes (assumed for heat exchanger tubeside pressure drop calculations), a constant of 23,000 should be used instead of 20,000.

For most common fluids the following equation also works quite well for ballpark checking and feasibility work:

$$W = 370\sqrt{\Delta P\,\rho d^5}$$

where W = flow in lb/hr
ΔP = friction loss in lb/in^2 (psi) per 100 ft
ρ = density in lb/ft^3
d = inside diameter of pipe in inches

This is a form of the Fanning or Darcy formula with friction factor = 0.0055. This friction factor corresponds to approximately the following:

Commercial steel pipes
Reynolds number = 10^5
Williams and Hazen C factor = 110

For other friction factors multiply the right hand side by

$$\sqrt{\frac{0.0055}{\text{friction factor}}}$$

The friction factor can be approximated by

Laminar flow,	$f = 16/Re$
Commercial pipes,	$f = 0.054/Re^{0.2}$
Smooth tubes,	$f = 0.046/Re^{0.2}$
Extremely rough pipes,	$f = 0.013$

where Re is the well known Reynolds number. In this case:

$$\text{friction factor} = \frac{h_L}{4\,(L/D)\,(u^2/2g)} \quad \text{(Fanning)}$$

where h_L = friction head loss in feet
L = length in feet
D = diameter in feet

Be careful when applying the friction factor. Sometimes it is defined as

$$\frac{h_L}{(L/D)\,(u^2/2g)} \quad \text{(Moody)}$$

The pressure drop equations apply to liquids. They also apply to compressible fluids for non-critical flow and $\Delta P < 10\%\ P_1$,

where ΔP = line pressure drop, psi
P_1 = upstream pressure, psia

For compressible flow where $\Delta P > 10\%\ P_1$, either break into sections where $\Delta P < 10\%\ P_1$ or use

$$\Delta P = P_1 - P_2 = \frac{2P_1}{P_1 + P_2} \left[0.323\left(\frac{fL}{d} + \frac{\ln(P_1/P_2)}{24}\right) S_1 U_1^2\right] \tag{1-4}$$

from Reference 2 that assumes isothermal flow of ideal gas. In Equation 1-4,

P_1, P_2 = upstream and downstream pressures in psi abs
S_1 = specific gravity of vapor relative to water = $0.00150\ MP_1/T$
d = pipe diameter, in.
U_1 = upstream velocity, ft/sec
f = friction factor (assume .005 for approximate work)
L = length of pipe, ft
ΔP = pressure drop in psi (rather than psi per standard length as before)
M = mol. wt.

Table 1-1 gives the equivalent length of straight pipe for various fittings (see pp. 10 and 11).

FLOW IN PARTIALLY FULL HORIZONTAL PIPES

The equations in the section "Piping Pressure Drop" are, of course, intended for use with full pipes. Reference 12 provides a rapid way to estimate whether a horizontal pipe carrying liquid is full. The criteria are

If $Q/d^{2.5} \geq 10.2$, the pipe is running full.

If $Q/d^{2.5} < 10.2$, Equation 1-5 is given for determining H/D, which is needed for a partially full flow analysis:

$$y_1 = 0.446 + 0.272x_1 + 0.0397x_1^2 - 0.0153x_1^3 - 0.003575x_1^4 \quad (1\text{-}5)$$

Once H/D is known, A (the flow area) can be found by

$$y_2 = -0.003597 + 0.3385x_2 + 1.3609x_2^2 - 0.9075x_2^3 \quad (1\text{-}6)$$

Also the "equivalent diameter" can be found by

$$y_3 = -0.01130 + 3.040x_2 - 3.461x_2^2 + 4.108x_2^3 - 2.638x_2^4 \quad (1\text{-}7)$$

This eliminates having to calculate the equivalent diameter by

$$D_e = 4\,(R_H) = 4\,(\text{cross-sectional flow area})/\text{wetted perimeter} \quad (1\text{-}8)$$

Note that for values of H/D greater than 0.5, but less than 1.0, $D_e/D > 1.0$. This results from the definition and is confirmed by my calculations and all of the references.

The equivalent diameter is used in place of the pipe diameter for non-circular ducts or partially full pipes. For example, it is used to calculate Re as a means of obtaining f. In determining:

$$Re = Du\rho/\mu$$

D_e is substituted for D; u is obtained by u = q/A.

Sometimes, however, A is expressed as $[\pi d^2]/[4(144)]$ with the $\pi/4(144)$ buried into an overall coefficient. For example, Crane[14] has a solved problem that uses the Darcy equation form:

$$Q = 19.65d^2 (h_L D/fL)^{0.5} \tag{1-9}$$

Here we could be confused with two diameter terms. Remember, however, that d is really a means of expressing area, so Crane[14] uses an "equivalent diameter of actual flow area" that is simply:

$$d = [144 (4) A/\pi]^{0.5} \tag{1-10}$$

Whereas D_e is substituted for D.

Nomenclature for Flow in Partially Full Horizontal Pipes

A = cross-sectional flow area (for partially full pipe), ft^2
D = pipe diameter, ft (also represents the quantity needing substitution by D_e)
d = pipe diameter, in. (also represents "equivalent diameter of actual flow area")
D_e = equivalent diameter = $4R_H$, ft
f = friction factor in $h_L = f_L u^2/D2g$
g = acceleration of gravity = 32.2 ft/sec^2
H = height of liquid in the pipe, ft
h_L = head loss, ft of flowing fluid
L = pipe length, ft
P = wetted perimeter, ft
Q = flow rate, gpm
q = flow rate, ft^3/sec
Re = Reynold's number, dimensionless
R_H = hydraulic radius = A/P, ft
u = velocity, ft/sec
x_1 = $\ln(Q/d^{2.5})$
x_2, y_1 = H/D
y_2 = A/D^2
y_3 = D_e/D
ρ = fluid density, lb/ft^3
μ = fluid viscosity, lb/ft-sec

Table 1-1
Equivalent Length of Valves and Fittings in Feet[9]

Nominal Pipe size in.	Globe valve or ball check valve	Angle valve	Swing check valve	Plug cock	Gate or ball valve	45° ell		Short rad. ell		Long rad. ell		Hard T.		Soft T.	
						Weld	thrd	Weld	thrd	Weld	thrd	Weld	thrd	Weld	thrd
1½	55	26	13	7	1	1	2	3	5	2	3	8	9	2	3
2	70	33	17	14	2	2	3	4	5	3	4	10	11	3	4
2½	80	40	20	11	2	2	..	5	..	3	..	12		3	..
3	100	50	25	17	2	2		6		4		14		4	
4	130	65	32	30	3	3		7		5		19		5	
6	200	100	48	70	4	4		11		8		28		8	
8	260	125	64	120	6	6		15		9		37		9	
10	330	160	80	170	7	7		18		12		47		12	
12	400	190	95	170	9	9		22		14		55		14	
14	450	210	105	80	10	10		26		16		62		16	
16	500	240	120	145	11	11		29		18		72		18	
18	550	280	140	160	12	12		33		20		82		20	
20	650	300	155	210	14	14		36		23		90		23	
22	688	335	170	225	15	15		40		25		100		25	
24	750	370	185	254	16	16		44		27		110		27	
30	—	—	—	312	21	21		55		40		140		40	
36	—	—	—		25	25		66		47		170		47	
42	—	—	—		30	30		77		55		200		55	
48	—	—	—		35	35		88		65		220		65	
54	—	—	—		40	40		99		70		250		70	
60	—	—	—		45	45		110		80		260		80	

90° miter bends			Enlargement					Contraction				
			Sudden			Std. red.		Sudden			Std. red.	
			Equiv. L in terms of small d									
2 miter	3 miter	4 miter	d/D = 1/4	d/D = 1/2	d/D = 3/4	d/D = 1/2	d/D = 3/4	d/D = 1/4	d/D = 1/2	d/D = 3/4	d/D = 1/2	d/D = 3/4
			5	3	1	4	1	3	2	1	1	—
			7	4	1	5	1	3	3	1	1	—
			8	5	2	6	2	4	3	2	2	—
			10	6	2	8	2	5	4	2	2	—
			12	8	3	10	3	6	5	3	3	—
			18	12	4	14	4	9	7	4	4	1
			25	16	5	19	5	12	9	5	5	2
			31	20	7	24	7	15	12	6	6	2
28	21	20	37	24	8	28	8	18	14	7	7	2
32	24	22	42	26	9	—	—	20	16	8	—	—
38	27	24	47	30	10	—	—	24	18	9	—	—
42	30	28	53	35	11	—	—	26	20	10	—	—
46	33	32	60	38	13	—	—	30	23	11	—	—
52	36	34	65	42	14	—	—	32	25	12	—	—
56	39	36	70	46	15	—	—	35	27	13	—	—
70	51	44										
84	60	52										
98	69	64										
112	81	72										
126	90	80										
190	99	92										

TWO-PHASE FLOW

Two-phase flow is beyond the scope of this pocket guide. One word of advice: Be careful when designing low pressure and/or flashing condensate lines. These deserve special care. Ruskin[10] has a quick method for condensate line sizing.

PIPING RULES OF THUMB

Tables 1-2, 1-3, and 1-4 give typical piping rules of thumb.

SONIC VELOCITY

To determine sonic velocity use

$$V_s = \sqrt{KgRT} \qquad (1\text{-}11)$$

where V_s = sonic velocity, ft/sec
K = C_P/C_v the ratio of specific heats at constant pressure to constant volume. This ratio is 1.4 for most diatomic gases.
g = 32.2 ft/sec^2
R = 1,544/mol wt
T = absolute temperature in °R

Table 1-2
Sizing Steam Piping in New Plants
Maximum Allowable Flow and Pressure Drop

	Laterals			Mains		
Pressure, psig	600	175	30	600	175	30
Density, lb/CF	0.91	0.41	0.106	0.91	0.41	0.106
ΔP, psi/100 ft	1.0	0.70	0.50	0.70	0.40	0.30
Nominal Pipe Size, in.	**Maximum lb/hr × 10^{-3}**					
3	7.5	3.6	1.2	6.2	2.7	0.9
4	15	7.5	3.2	12	5.7	2.5
6	40	21	8.5	33	16	6.6
8	76	42	18	63	32	14
10	130	76	32	108	58	25
12	190	115	50	158	87	39
14	260	155	70	217	117	54
16	360	220	100	300	166	78
18	...	300	130	...	227	101
20	...	...	170	...	...	132

Note:

1. *600 psig steam is at 750°F, 175 psig and 30 psig are saturated.*
2. *On 600 psig flow ratings, internal pipe sizes for larger nominal diameters were taken as follows: 18/16.5 in., 14/12.8 in., 12/11.6 in., 10/9.75 in.*
3. *If other actual I.D. pipe sizes are used, or if local superheat exists on 175 psig or 30 psig systems, the* allowable pressure drop *shall be the governing design criterion.*

Table 1-3
Sizing Cooling Water Piping in New Plants
Maximum Allowable Flow, Velocity, and Pressure Drop

	Laterals			Mains		
Pipe Size in.	Flow gpm	Vel. ft/sec	ΔP ft/100 ft	Flow gpm	Vel. ft/sec	ΔP ft/100 ft
3	100	4.34	4.47	70	3.04	2.31
4	200	5.05	4.29	140	3.53	2.22
6	500	5.56	3.19	380	4.22	1.92
8	900	5.77	2.48	650	4.17	1.36
10	1,500	6.10	2.11	1,100	4.48	1.19
12	2,400	6.81	2.10	1,800	5.11	1.23
14	3,100	7.20	2.10	2,200	5.13	1.14
16	4,500	7.91	2.09	3,300	5.90	1.16
18	6,000	8.31	1.99	4,500	6.23	1.17
20	....	....	....	6,000	6.67	1.17
24	....	....	....	11,000	7.82	1.19
30	....	....	....	19,000	8.67	1.11

To determine the critical pressure ratio for gas sonic velocity across a nozzle or orifice use

$$\text{critical pressure ratio} = [2/(K + 1)]^{k/(k-1)} \qquad (1\text{-}12)$$

If pressure drop is high enough to exceed the critical ratio, sonic velocity will be reached. When K = 1.4, ratio = 0.53.

Table 1-4
Sizing Piping for Miscellaneous Fluids

Dry Gas	100 ft/sec
Wet Gas	60 ft/sec
High Pressure Steam	150 ft/sec
Low Pressure Steam	100 ft/sec
Air	100 ft/sec
Vapor Lines General	Max. velocity 0.3 mach 0.5 psi/100 ft
Light Volatile Liquid Near Bubble Pt. Pump Suction	0.5 ft head total suction line
Pump Discharge, Tower Reflux	3–5 psi/100 ft
Hot Oil headers	1.5 psi/100 ft
Vacuum Vapor Lines below 50 MM Absolute Pressure	Allow max. of 5% absolute pressure for friction loss

CONTROL VALVE DESIGN

Notes:

1. References 4 and 5 were used extensively for this section. The sizing procedure is generally that of Fisher Controls Company.
2. Use manufacturers' data where available. This handbook will provide approximate parameters applicable to a wide range of manufacturers.

3. For any control valve design, be sure to use one of the modern methods, such as that given here, that takes into account such things as control valve pressure recovery factors and gas transition to incompressible flow at critical pressure drop.

Liquid Flow

Recall the previous discussion of converting PV to $u^2/2g$. Across a control valve, the fluid is accelerated to some maximum velocity. At this point the pressure reduces to its lowest value. If this pressure is lower than the liquid's vapor pressure, flashing will produce bubbles or cavities of vapor. The pressure will rise or "recover" downstream of the lowest pressure point. If the pressure rises to above the vapor pressure, the bubbles or cavities collapse. This causes noise, vibration, and physical damage.

When there is a choice, design for no flashing. When there is no choice, locate the valve to flash into a vessel if possible. If flashing or cavitation cannot be avoided, select hardware that can withstand these severe conditions. The downstream line will have to be sized for two-phase flow. It is a good idea to use a long conical adaptor from the control valve to the downstream line.

When sizing liquid control valves first use

$$\Delta P_{allow} = K_m(P_1 - r_c P_v) \qquad (1\text{-}13)$$

where ΔP_{allow} = maximum allowable differential pressure for sizing purposes, psi

K_m = valve recovery coefficient (see Table 1-7)
r_c = critical pressure ratio (see Figures 1-1 and 1-2)
P_1 = body inlet pressure, psia
P_v = vapor pressure of liquid at body inlet temperature, psia

This gives the maximum ΔP that is effective in producing flow. Above this ΔP, no additional flow will be produced

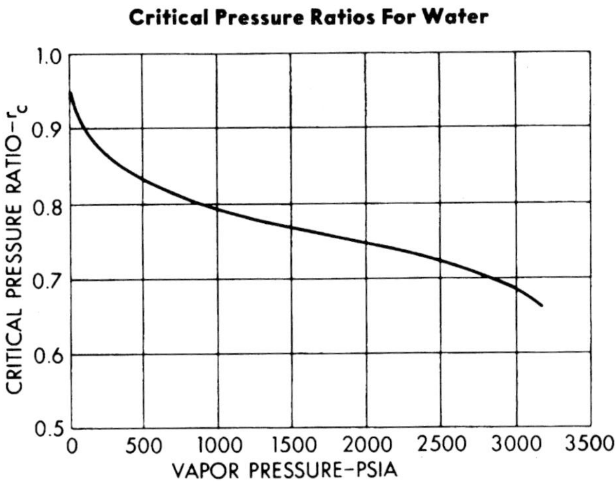

Figure 1-1. Enter on the abscissa at the water vapor pressure at the valve inlet. Proceed vertically to intersect the curve. Move horizontally to the left to read rc on the ordinate.[4]

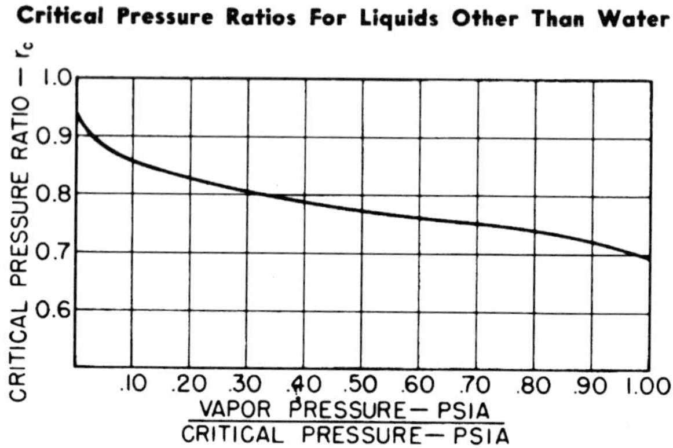

Figure 1-2. Determine the vapor pressure/critical pressure ratio by dividing the liquid vapor pressure at the valve inlet by the critical pressure of the liquid. Enter on the abscissa at the ratio just calculated and proceed vertically to intersect the curve. Move horizontally to the left and read r_c on the ordinate.[4]

because flow will be restricted by flashing. Do not use a number higher than ΔP_{allow} in the liquid sizing formula. Some designers use as the minimum pressure for flash check the upstream absolute pressure minus two times control valve pressure drop.

Table 1-5 gives critical pressures for miscellaneous fluids. Table 1-6 gives relative flow capacities of various types of control valves. This is a rough guide to use in lieu of manufacturer's data.

Table 1-5
Critical Pressure of Various Fluids, psia*

Ammonia	1636	Isobutane	529.2
Argon	705.6	Isobutylene	580
Butane	550.4	Methane	673.3
Carbon Dioxide	1071.6	Nitrogen	492.4
Carbon Monoxide	507.5	Nitrous Oxide	1047.6
Chlorine	1118.7	Oxygen	736.5
Dowtherm A	465	Phosgene	823.2
Ethane	708	Propane	617.4
Ethylene	735	Propylene	670.3
Fluorine	808.5	Refrigerant 11	635
Helium	33.2	Refrigerant 12	596.9
Hydrogen	188.2	Refrigerant 22	716
Hydrogen Chloride	1198	Water	3206.2

**For values not listed, consult an appropriate reference book.*

The liquid sizing formula is

$$C_v = Q\sqrt{\frac{G}{\Delta P}} \tag{1-14}$$

where C_v = liquid sizing coefficient
Q = flow rate in gpm
ΔP = body differential pressure, psi
G = specific gravity (water at 60°F = 1.0)

Table 1-6
Relative Flow Capacities of Control Valves[5, 20]

Valve Type	C_d*	C_dF_p†	C_dF_L**
Double-seat globe	12	11	11
Single-seat top-guided globe	11.5	10.8	10
Single-seat split body	12	11.3	10
Sliding gate	6–12	6–11	na
Single-seat top-entry cage	13.5	12.5	11.5
Eccentric rotating plug (Camflex)	14	13	12
60° open butterfly	18	15.5	12
Single-seat Y valve (300 & 600 lb)	19	16.5	14
Saunders type (unlined)	20	17	na
Saunders type (lined)	15	13.5	na
Throttling (characterized) ball	25	20	15
Single-seat streamlined angle (flow-to-close)	26	20	13
90° open butterfly (average)	32	21.5	18

Note: This table may serve only as a rough guide because actual flow capacities differ between manufacturer's products and individual valve sizes.[20]

**Valve flow coefficient $C_v = C_d \times d^2$ (d = valve dia., in.)*

†C_v/d^2 of valve when installed between pipe reducers (pipe dia. 2 × valve dia.)

***C_v/d^2 of valve when undergoing critical (choked) flow conditions*

Two liquid-control-valve-sizing rules of thumb are

1. No viscosity correction necessary if viscosity ≤ 20 centistokes.
2. For sizing a flashing control valve, add the C_v's of the liquid and the vapor.

GAS AND STEAM FLOW

The gas and steam sizing formulas are

Gas

$$C_g = \frac{Q}{\sqrt{\frac{520}{GT}}\, P_1 \sin\left[\frac{3417}{C_1}\sqrt{\frac{\Delta P}{P_1}}\right]_{deg}} \tag{1-15}$$

Steam (under 1,000 psig)

$$C_s = \frac{Q_s\,(1 + 0.00065\,T_{sh})}{P_1 \sin\left[\frac{3417}{C_1}\sqrt{\frac{\Delta P}{P_1}}\right]_{deg}} \tag{1-16}$$

Steam and vapors (all vapors, including steam under any pressure conditions)

$$C_g = \frac{Q_s}{1.06\sqrt{d_1 P_1}\, \sin\left[\frac{3,417}{C_1}\sqrt{\frac{\Delta P}{P_1}}\right]_{deg}} \tag{1-17}$$

When the bracketed quantity in the equations equals or exceeds 90°, critical flow is indicated. The quantity must be limited to 90°. This then becomes unity because sin 90° = 1.

Explanation of terms:

C_1 = C_g/C_v (some sizing methods use C_f or Y in place of C_1)
C_g = gas sizing coefficient
C_s = steam sizing coefficient
C_v = liquid sizing coefficient
d_1 = density of steam or vapor at inlet, lb/ft^3
G = gas specific gravity = mol wt/29
P_1 = valve inlet pressure, psia
ΔP = pressure drop across valve, psi
Q = gas flow rate, scfh
Q_s = steam or vapor flow rate, lb/hr
T = absolute temperature of gas at inlet, °R
T_{sh} = degrees of superheat, °F

The control valve coefficients in Table 1-8 are for full open conditions. The control valve must be designed to operate at partial open conditions for good control. Figure 1-3 shows partial open performance for a number of trim types.

CONTROL VALVE RULES OF THUMB

1. Design tolerance. Many use the greater of the following:

$$Q_{sizing} = 1.3\ Q_{normal}$$
$$Q_{sizing} = 1.1\ Q_{maximum}$$

(text continued on page 25)

Table 1-7
Average Valve-Recovery Coefficients, K_m and C_l*[5]

Type of Valve	K_m	C_l
Cage-trim globes:		
Unbalanced	0.80	33
Balanced	0.70	33
Butterfly:		
Fishtail	0.43	16
Conventional	0.55	24.7
Ball:		
Vee-ball, modified-ball, etc.	0.40	22
Full-area ball	0.30	
Conventional globe:		
Single and double port (full port)	0.75	35
Single and double port (reduced port)	0.65	35
Three way	0.75	
Angle:		
Flow tends to open (standard body)	0.85	
Flow tends to close (standard body)	0.50	
Flow tends to close (venturi outlet)	0.20	
Camflex:		
Flow tends to close	0.72	24.9
Flow tends to open	0.46	31.1
Split body	0.80	35

**For use only if not available from manufacturer.*

Table 1-8
Correlations of Control Valve Coefficients[5]

$$C_l = 36.59\ C_f \qquad K_m = C_f^2 = F_L^{\ 2}$$
$$C_l = 36.59\ \sqrt{K_m} \qquad C_s = 1.83\ C_f\ C_v$$
$$C_g = C_l\ C_v \qquad C_v = 19.99\ C_s/C_l$$
$$Cg = 19.99\ C_s$$

Values of K_m calculated from C_f agree within 10% of published data of K_m.
Values of C_l calculated from K_m are within 21% of published data of C_l.

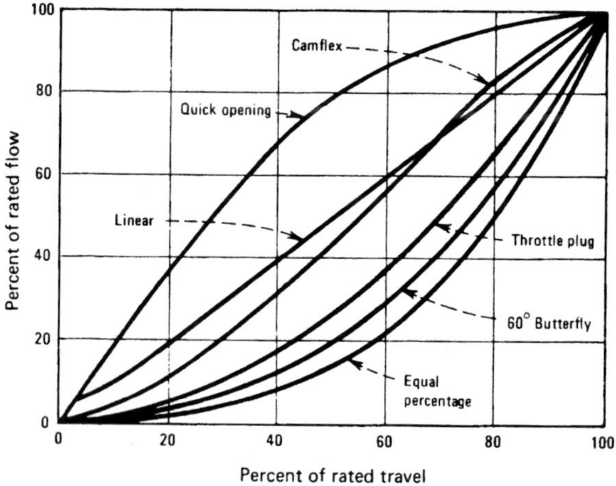

Figure 1-3. These are characteristic curves of common valves.[5]

2. Type of trim. Use equal percentage whenever there is a large design uncertainty or wide rangeability is desired. Use linear for small uncertainty cases.

   Limit max/min flow to about 10 for equal percentage trim and 5 for linear. Equal percentage trim usually requires one larger nominal body size than linear.
3. For good control where possible, make the control valve take 50%–60% of the system flowing head loss.
4. For saturated steam, keep control valve outlet velocity below 0.25 mach.
5. Keep valve inlet velocity below 300 ft/sec for 2 in. and smaller, and 200 ft/sec for larger sizes.

SAFETY RELIEF VALVE DESIGN

The ASME code provides the basic requirements for overpressure protection. Section I, *Power Boilers,* covers fired and unfired steam boilers. All other vessels including exchanger shells and similar pressure-containing equipment fall under Section VIII, *Pressure Vessels.* API RP 520 and lesser API documents supplement the ASME code. These codes specify allowable accumulation, which is the difference between relieving pressure at which the valve reaches full rated flow and set pressure at which the valve starts to open. Accumulation is expressed as percentage of set pressure in Table 1-9.

Table 1-9
Accumulation Expressed as Percentage of Set Pressure

	ASME Section I Power Boilers	ASME Section VIII Pressure Vessels	Typical Design for Compressors Pumps and Piping
Liquids			
Thermal expansion	. . .	10	25
Fire	. . .	20	20
Steam			
Overpressure	3	10	10
Fire	. . .	20	20
Gas or vapor			
Overpressure	. . .	10	10
Fire	. . .	20	20

Full liquid containers require protection from thermal expansion. Such relief valves are generally quite small. Three examples are

1. Cooling water than can be blocked in with hot fluid still flowing on the other side of an exchanger.
2. Long lines to tank farms that can lie stagnant exposed to the sun.
3. Plant piping that can be blocked in with steam tracing on.

Sizing

Use manufacturer's sizing charts and data where available. In lieu of manufacturer's data use the formula

$$u = 0.4\sqrt{2g\Delta h} \tag{1-18}$$

where Δh = head loss in feet of flowing fluid
u = velocity in ft/sec
$g = 32.2$ ft/sec^2

This will give a conservative relief valve area. For compressible fluids, use Δh corresponding to ½ P_1 if head difference is greater than that corresponding to ½ P_1 (because sonic velocity occurs). If head difference is below that corresponding to ½ P_1, use actual Δh.

For vessels filled with only gas or vapor and exposed to fire use[8]

$$A = \frac{0.042\,A_s}{\sqrt{P_1}} \tag{1-19}$$

A = calculated nozzle area, in.^{2}
P_1 = set pressure (psig) × (1 + fraction accumulation) + atmospheric pressure, psia. For example, if accumulation = 10%, then (1 + fraction accumulation) = 1.10
A_s = exposed surface of vessel, ft^2

This will also give conservative results. For heat input from fire to liquid-containing vessels, see "Determination of Rates of Discharge" in this chapter.

The set pressure of a conventional valve is affected by back pressure. The spring setting can be adjusted to compensate for *constant* back pressure. For a variable back pressure of greater than 10% of the set pressure, it is customary to go to the balanced bellows type, which can generally tolerate variable back pressure of up to 40% of set pressure. Table 1-10 gives standard orifice sizes.

Determination of Rates of Discharge

The more common causes of overpressure are

1. External fire
2. Heat exchanger tube failure
3. Liquid expansion
4. Cooling water failure
5. Electricity failure
6. Blocked outlet
7. Failure of automatic controls
8. Loss of reflux
9. Chemical reaction (this heat can sometimes exceed the heat of an external fire)

Plants, situations, and causes of overpressure tend to be dissimilar enough to discourage preparation of generalized calculation procedures for the rate of discharge. In lieu of a set procedure, most of these problems can be solved satisfactorily by conservative simplification and analysis. It should be noted also that, by general assumption, two unrelated emergency conditions will not occur simultaneously.

Table 1-10
Standard Orifice Sizes for Flanged Steel Safety Relief Valves

Standard Orifice Designation	Orifice Area (in.^{2})	1 × 2	1.5 × 2	1.5 × 2.5	1.5 × 3	2 × 3	2.5 × 4	3 × 4	4 × 6	6 × 8	6 × 10	8 × 10
D	0.110	•	•	•								
E	0.196	•	•	•								
F	0.307	•	•	•								
G	0.503			•	•	•						
H	0.785				•	•						
J	1.287					•	•	•				
K	1.838							•				
L	2.853							•	•			
M	3.60								•			
N	4.34								•			
P	6.38								•			
Q	11.05									•		
R	16.0									•	•	
T	26.0											•

Valve Body Size (Inlet Diameter × Outlet Diameter), in.

The first three causes of overpressure on the list are more amenable to generalization than the others and will be discussed.

Fire

The heat input from fire is discussed in API RP 520.[8] One form of their equation for liquid-containing vessels is

$$Q = 21{,}000\ F A_w^{0.82} \tag{1-20}$$

where Q = heat absorption, Btu/hr
A_w = total wetted surface, ft^2
F = environmental factor

The environmental factors represented by F are

Bare vessel = 1.0
Insulated = 0.3/insulation thickness, in.
Underground storage = 0.0
Earth covered above grade = 0.03

The height above grade for calculating wetted surface should be

1. For vertical vessels—at least 25 ft above grade or other level at which a fire could be sustained.
2. For horizontal vessels—at least equal to the maximum diameter.
3. For spheres or spheroids—whichever is greater, the equator or 25 ft.

Heat Exchanger Tube Failure

1. Use the fluid entering from twice the cross section of one tube as stated in API RP 520[8] (one tube cut in half exposes two cross sections at the cut).
2. Use the old standby, $\Delta h = u^2/2g$, to calculate leakage. Because this acts similar to an orifice, we need a coefficient; use 0.7. So,

$$u = 0.7\sqrt{2g\Delta h} \tag{1-21}$$

For compressible fluids, if the downstream head is less than ½ the upstream head, use ½ the upstream head as Δh. Otherwise, use the actual Δh.

Liquid Expansion

Equation 1-22 can be used for sizing relief valves for liquid expansion.[8]

$$Q = \frac{BH}{500\ GC} \tag{1-22}$$

where Q = required capacity, gpm
H = heat input, Btu/hr
B = coefficient of volumetric expansion per °F:
= 0.0001 for water
= 0.0010 for light hydrocarbons
= 0.0008 for gasoline
= 0.0006 for distillates
= 0.0004 for residual fuel oil
G = specific gravity
C = specific heat, Btu/lb °F

Rules of Thumb for Safety Relief Valves

1. Check metallurgy for light hydrocarbons flashing during relief. Very low temperatures can be produced.
2. Always check for reaction force from the tailpipe.
3. Hand jacks are a big help on large relief valves for several reasons. One is to give the operator a chance to reseat a leaking relief valve.
4. Flat seated valves have an advantage over bevel seated valves if the plant forces have to reface the surfaces (usually happens at midnight).
5. The maximum pressure from an explosion of a hydrocarbon and air is 7 times initial pressure, unless it occurs in a long pipe where a standing wave can be set up. It may be cheaper to design some small vessels to withstand an explosion than to provide a safety relief system. It is typical to specify ¼ in. as minimum plate thickness (for carbon steel only).

RELIEF MANIFOLDS

Mak[15] has developed an improved method of relief valve manifold design. The API[8] has adopted this method, which starts at the flare tip (atmospheric pressure) and calculates backwards to the relief valves, thus avoiding the trial and error of other methods. This is especially helpful when a large number of relief valves may discharge simultaneously to the same manifold.

Mak's developed isothermal equation (based on the manifold outlet pressure rather than the inlet) is:

$$fL/D = (1/M_2^2)(P_1/P_2)^2[1 - (P_2/P_1)^2] - \ln(P_1/P_2)^2 \quad (1\text{-}23)$$

where D = header diameter, ft
f = Moody friction factor
L = header equivalent length, ft
M_2 = Mach number at the header outlet
P_1, P_2 = inlet and outlet header pressures, psia

The equation for M_2 is as follows:

$$M_2 = 1.702 \times 10^{-5} [W/(P_2D^2)](ZT/M_W)^{1/2} \quad (1\text{-}24)$$

where W = gas flow rate, lb/hr
Z = gas compressibility factor
T = absolute temperature, °R
M_W = gas molecular weight

To simplify and speed calculations, Mak provides Figure 1-4. The method is applied as follows:

1. Assume diameters of all pipes in the network.
2. Starting at the flare tip, calculate logical segments using Figure 1-4 until all relief valve outlet pressures are found.
3. Check all relief valves against their MABP.
   Case A. The calculated back pressure at the lowest set relief valve on a header is much smaller than its MABP. Reduce header size.

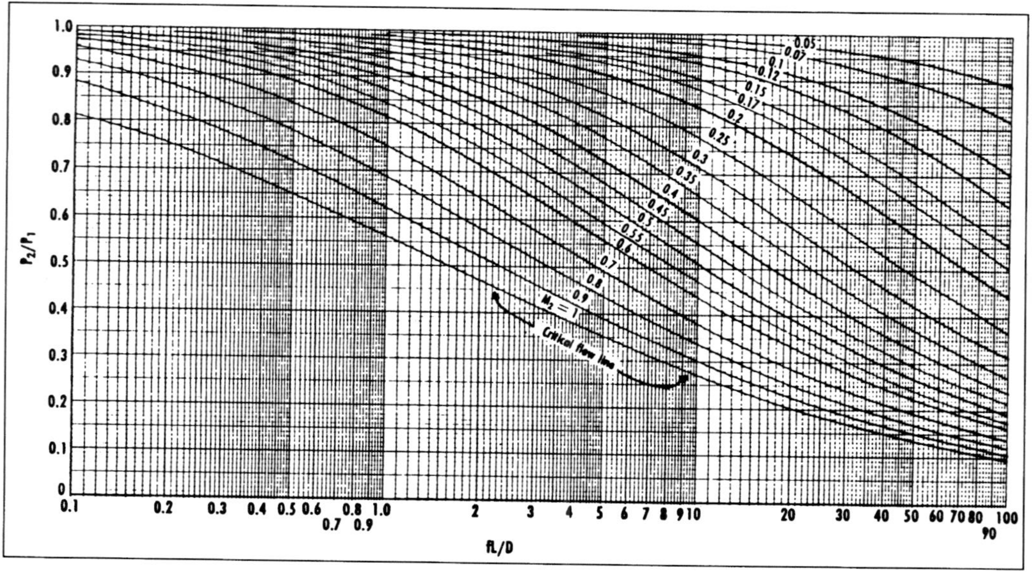

Figure 1-4. Isothermal flow chart.

Case B. The calculated back pressure at the lowest set relief valve on a header is close to and below its MABP. The header size is correct.

Case C. The calculated back pressure at the lowest set relief valve on a header is above its MABP. Increase header size.

4. Use judgment in attempting to optimize. Try to preferentially reduce the sizes of the longest runs or those having the most fittings.

Figure 1-5 and Table 1-11 illustrate a sample problem (Z, the compressibility factor, is assumed to be 1.0).

Often, if both high and low pressure relief valves need to relieve simultaneously, parallel high and low pressure headers terminating at the flare knockout drum are the economical choice. Be sure to check for critical flow at key points in the high pressure header.

$$P_{crit} = (W/408d^2)(ZT/M_W)^{0.5}$$

where P_{crit} = critical pressure, psia
W = gas flow rate, lb/hr
d = pipe ID, in.
Z = gas compressibility factor
T = gas temperature, °R
M_W = gas molecular weight

(text continued on page 38)

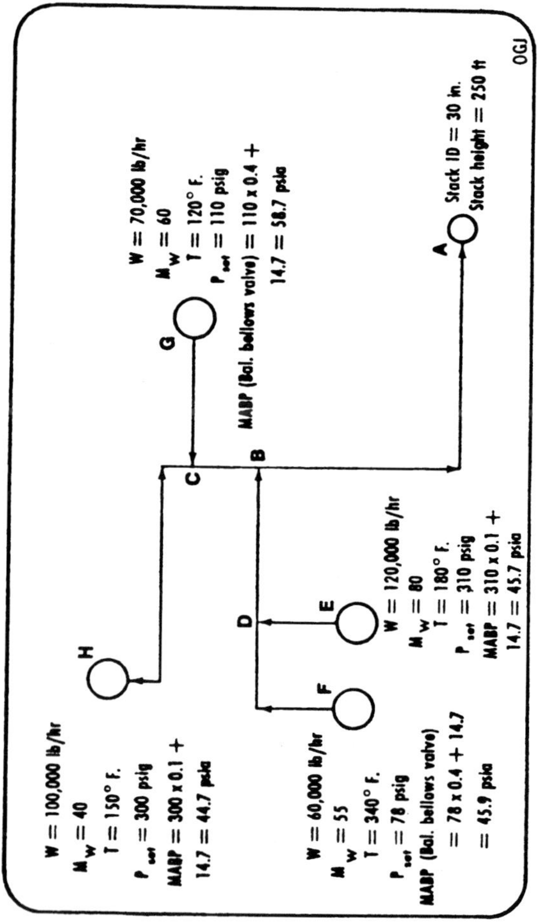

Figure 1-5. Example of relief/flare system.

Table 1-11
Back-Pressure Calculations for Example

	Line Segment							
Line	Stack	AB	BD	DF	DE	BC	CH	CG
Size	30 in. (std)	18 in. (std)	12 in. (std)	8 in. (std)	8 in. (std)	12 in. (std)	10 in. (std)	6 in. (std)
D, ft	2.5	1.44	1	0.665	0.665	1	0.835	0.505
A, ft^2	4.91	1.623	0.785	0.347	0.347	0.785	0.548	0.201
L, ft	250	1000	200	180	100	115	300	150
f	0.011	0.012	0.013	0.014	0.014	0.013	0.0135	0.015
W, lb/hr	350,000	350,000	180,000	60,000	120,000	170,000	100,000	70,000
M_W	56	56	69.5	55	80	46.4	40	60
T, °F.	186.6	186.8	233.3	340	180	137.6	150	120
P_2, psia	P_{atm} = 14.7	P_A = 15.12	P_B = 34.43	P_D = 38.04	P_D = 38.04	P_B = 34.43	P_C = 37.02	P_C = 37.02
fL/D	1.10	8.35	2.60	3.79	2.10	1.50	4.85	4.45
M_2	0.220	0.646	0.281	0.232	0.344	0.302	0.257	0.391
From Figure 1-4	P_{atm}/P_A = 0.972	P_A/P_B = 0.440	P_B/P_D = 0.905	P_D/P_F = 0.910	P_D/P_E = 0.890	P_B/P_C = 0.930	P_C/P_H = 0.867	P_C/P_G = 0.755
P_1, psia	P_A = 15.12	P_B = 34.43	P_D = 38.04	P_F = 41.80	P_E = 42.74	P_C = 37.02	P_H = 42.70	P_G = 49.03

(text continued from page 35)

The check for critical pressure has two purposes:

1. If, for a segment, the P_2 calculated is less than P_{crit} then the flow is critical and P_{crit} is used in place of P_2.
2. The main flare header should not be designed for critical flow at the entrance to the flare stack, or else noise and vibration will result

A reading below $M_2 = 1$ on Figure 1-4 also indicates critical flow. In such a case, read the graph at $M_2 = 1$.

Mak's article shows how the isothermal assumption is slightly conservative (higher relief valve outlet pressures) when compared to adiabatic.

METERING

Orifice

$$(U_o^2 - U_p^2)^{1/2} = C_o\,(2g\Delta h)^{1/2} \qquad (1\text{-}25)$$

Permanent head loss % of Δh

D_o/D_p	Permanent Loss
0.2	95
0.4	82
0.6	63
0.8	40

One designer uses permanent loss = $\Delta h\ (1 - C_o)$

where U_o = velocity through orifice, ft/sec
U_p = velocity through pipe, ft/sec
$2g$ = 64.4 ft/sec^2
Δh = orifice pressure drop, ft of fluid
D = diameter
C_o = coefficient. (Use 0.60 for typical application where D_o/D_p is between 0.2 and 0.8 and Re at vena contracta is above 15,000.)

Venturi

Same equation as for orifice:

$$C_o = 0.98 \qquad (1\text{-}26)$$

Permanent head loss approximately 3–4% Δh.

Rectangular Weir

$$F_v = 3.33\ (L - 0.2H)\ H^{3/2} \qquad (1\text{-}27)$$

where F_v = flow in ft^3/sec
L = width of weir, ft
H = height of liquid over weir, ft

Pitot Tube

$$\Delta h = u^2/2g$$

REFERENCES

1. *Cameron Hydraulic Data,* 13th ed., Ingersoll-Rand Company, Cameron Pump Division, 1965.
2. Maxwell, J. B., *Data Book on Hydrocarbons,* Van Nostrand, 1965.
3. Perry and Chilton, *Chemical Engineers' Handbook,* McGraw-Hill Inc., 1973.
4. Fisher Controls Company, *Sizing and Selection Data,* Catalog 10.
5. Chalfin, Fluor Corp., "Specifying Control Valves," *Chemical Engineering,* McGraw-Hill, Inc., Oct. 14, 1974. Used with permission.
6. Rearick, "How to Design Pressure Relief Systems," Parts I and II, *Hydrocarbon Processing,* Aug./Sept. 1969.
7. ASME Boiler and Pressure Vessel Code Sections I and VIII.
8. *Recommended Practice for the Design and Installation of Pressure Relieving Systems in Refineries,* Part I: "Design," latest edition, Part II: "Installation," latest edition. RP 520 American Petroleum Institute.
9. *NGPSA Engineering Data Book,* 9th ed., Natural Gas Processors Suppliers Association, 1972.

10. Ruskin, Richard P., "Calculating Line Sizes for Flashing Steam-Condensate," *Chemical Engineering,* Aug. 18, 1975.
11. Isaacs, Marx, "Pressure Relief Systems," *Chemical Engineering,* Feb. 22, 1971.
12. Durand, A. A. and M. Marquez-Lucero, "Determining Sealing Flow Rates in Horizontal Run Pipes," *Chemical Engineering,* March 1998, p. 129.
13. King, H. W. and E. F. Brater, *Handbook of Hydraulics,* 4th Spanish Edition, McGraw-Hill, New York, 1962, Chap. 8, pp. 347–351.
14. "Flow of Fluids through Valves, Fittings and Pipe," Crane Co. Technical Paper 410.
15. Mak, "New Method Speeds Pressure-Relief Manifold Design," *Oil and Gas Journal,* Nov. 20, 1978.
16. Crocker, Sabin, *Piping Handbook,* McGraw-Hill, Inc., 1945.
17. Standing, M. B. and D. L. Katz, Trans., AIME 146, 159 (1942).
18. Branan, C. R., *Rules of Thumb for Chemical Engineers,* 2nd ed, Gulf Publishing Co., Houston, TX, 1998.
19. Branan, C. R., "Estimating Pressure Drop," *Chemical Engineering,* Aug. 28, 1978. Note: Equations were generated using FLEXCURV, Gulf Publishing Co.
20. *Handbook of Control Valves,* Instrument Society of America, p. 17.

2
Pumps and Motors

This chapter will not contain detailed piping procedures for pumps. There are many good sources for this type information such as References 3, 4, 5, 6, 7, and 8. Suction piping, especially for liquids near their boiling point, requires very accurate design, which is beyond the scope of this small handbook. Instead, helpful quick methods will be given to estimate pump and motor size.

AFFINITY (OR FAN) LAWS

Dynamic type pumps obey the Affinity Laws:

1. Capacity varies directly with impeller diameter and speed.
2. Head varies as the square of impeller diameter and speed.
3. Horsepower varies as the cube of impeller diameter and speed.

PUMP HORSEPOWER

The handiest pump horsepower formula for a process engineer is

$$HP = \frac{GPM\,(\Delta P)}{1715\,(Eff.)} \tag{2-1}$$

where HP = pump horsepower
GPM = gallons per minute
ΔP = delivered pressure (discharge minus suction), psi
Eff. = pump efficiency

PUMP EFFICIENCY

An equation was developed by the author from the pump efficiency curves presented in Reference 1 and provided by the M. W. Kellogg Co. The curves were found to check vendor data well. The equation is admittedly bulky appearing, but is easier to use than it appears.

$$\begin{aligned} \text{Efficiency} = {} & 80 - 0.2855F + 3.78 \times 10^{-4}\,FG - 2.38 \\ & \times 10^{-7}\,FG^2 + 5.39 \times 10^{-4}\,F^2 - 6.39 \\ & \times 10^{-7}\,F^2G + 4 \times 10^{-10}\,F^2G^2 \end{aligned} \tag{2-2}$$

where Efficiency = pump percentage efficiency
F = developed head, ft
G = flow, gpm

Ranges of applicability:

F = 50–300 ft
G = 100–1000 gpm

Equation 2-2 gives results within about 7% of the aforementioned pump curves. This means within 7% of the curve valve, not 7% absolute; that is, if the curve valve is 50%, the equation will be within the range 50 ± 3.5%.

For flows in the range 25–99 gpm, a rough efficiency can be obtained by using Equation 2-2 for 100 gpm and then subtracting 0.35%/gpm times the difference between 100 gpm and the low flow gpm. For flows at the bottom of the range (25–30 gpm), this will give results within about 15% for the middle of the head range and 25% at the extremes. This is adequate for ballpark estimates at these low flows. The horsepower at the 25–30 gpm level is generally below 10.

MOTOR EFFICIENCY

Table 2-1 yields ballpark motor efficiencies. The table applies to three-phase induction motors only. For single-phase, multiply by about 0.8. Single-phase motors are not generally recommended above 10 HP.

Table 2-2 contains handy data for determining starter size.

Table 2-1
Motor Efficiencies

Standard Motor Horsepower Ratings	Efficiency	Standard Motor Horsepower Ratings	Efficiency
1	80	250	90
2	82	500	93
3	84	600	93
5	85	700	93
7½	85	800	94
10	85	900	94
15	86	1,000	94
20	87	1,250	94
25	88	1,500	94
30	89	1,750	94
40	89	2,000	94
50	89	2,250	94
75	90	2,500	95
100	90	3,000	95
125	90	3,500	95
150	90	4,000	95
200	90	4,500	95
		5,000	95

Table 2-2
Starter Size Versus Horsepower

	Starter Size	00	0	1	2	3	4	5
Three-Phase	208–220 volt	1½	3	7½	15	30	50	100
	440–550 volt	2	5	10	25	50	100	200
Single-Phase	110 volt	⅓	1	2	3	...	...	...
	220 volt	1	2	3	5	...	...	...

REFERENCES

1. *NGPSA Engineering Data Book,* Natural Gas Processors Suppliers Association, 8th ed., 1966 and 9th ed., 1972.
2. *Motor/Price Selector,* Reliance Electric Company, Bulletin B-111, Industrial Drives Group, Cleveland, Ohio.
3. Kern, Robert, "How to Design Piping For Pump-Suction Conditions," *Chemical Engineering,* April 28, 1975.
4. Van Blarcom, Peter P., Yarway Corp., "Bypass Systems for Centrifugal Pumps, *Chemical Engineering,* Feb. 4, 1974.
5. Neerken, Richard F., The Ralph M. Parsons Co., "Pump Selection for the Chemical Process Industries," *Chemical Engineering,* Feb. 18, 1974.
6. Selection of Boiler Feed Pumps, *Bulletin 781.1,* Goulds Pumps, Inc. Seneca Falls, N.Y., Jan. 31, 1968.
7. Caplan, F., "Estimated Minimum Required Flows through Pumps," *Chemical Engineering,* March 17, 1975.
8. Kern , Robert, "Use Nomographs to Quickly Size Pump Piping and Components," *Hydrocarbon Processing,* March 1973.

3
Compressors and Steam Turbines

COMPRESSORS

This chapter discusses centrifugal compressors and steam turbines. Centrifugal compressors are best discussed by going through a calculation procedure and discussing each part.

First, the required head is calculated. Either the polytropic or adiabatic head can be used to calculate horsepower as long as the polytropic or adiabatic efficiency is used with the companion head.

Polytropic Head

$$H_{poly} = \frac{ZRT_1}{(N-1)/N}\left[\left(\frac{P_2}{P_1}\right)^{(N-1)/N} - 1\right] \qquad (3\text{-}1)$$

Adiabatic Head

$$H_{AD} = \frac{ZRT_1}{(K-1)/K}\left[\left(\frac{P_2}{P_1}\right)^{(K-1)/K} - 1\right] \qquad (3\text{-}2)$$

where Z = average compressibility factor; using 1.0 will yield conservative results

R = 1,544/mol. wt.

T_1 = suction temperature, °R

P_1, P_2 = suction, discharge pressures, psia

K = adiabatic exponent, C_p/C_v

N = polytropic exponent, $\frac{N-1}{N} = \frac{K-1}{KE_p}$

E_p = polytropic efficiency, use 75% for preliminary work

E_A = adiabatic efficiency

The polytropic and adiabatic efficiencies are related as follows:

$$E_A = \left[\frac{\left(\frac{P_2}{P_1}\right)^{(K-1)/K} - 1}{\left(\frac{P_2}{P_1}\right)^{(N-1)/N} - 1}\right] = \left[\frac{\left(\frac{P_2}{P_1}\right)^{(K-1)/K} - 1}{\left(\frac{P_2}{P_1}\right)^{(K-1)/KE_p} - 1}\right] \qquad (3\text{-}3)$$

The gas horsepower is calculated using the companion head and efficiency.

From Polytropic Head

$$HP = \frac{W\,H_{poly}}{E_p 33,000} \tag{3-4}$$

From Adiabatic Head

$$HP = \frac{W\,H_{AD}}{E_A 33,000} \tag{3-5}$$

where HP = gas horsepower
W = flow, lb/min

To the gas horsepower is added bearing and oil seal losses. Use 50 hp in lieu of manufacturer's data for large machines.

The discharge temperature is calculated as follows:

$$t_2 = \frac{H_{poly}}{ZR\left(\frac{K}{K-1}\right)E_p} + t_1 \tag{3-6}$$

Often the temperature of the gas must be limited. Sometimes temperature is limited to protect against polymerization as in olefin or butadiene plants. At temperatures greater than 450–500°F, the approximate mechanical limit, problems of sealing and casing growth start to occur. High temperature requires a special and more costly machine. Most multistage applications are designed to stay below 250–300°F.

Intercooling can be used to hold desired temperatures for high overall compression ratio applications. This can be done between stages in a single compressor frame or between series frames. Sometimes economics rather than a temperature limit dictate intercooling.

Sometimes for high compression ratio applications, the job cannot be done in a single compressor frame. Usually a frame will not contain more than about eight stages (wheels). There is a maximum head that one stage can handle. This depends upon the gas properties and inlet temperature. Usually this will run 7,000 to 11,000 feet for a single stage. In lieu of manufacturer's data use eight maximum stages per frame. Then subtract one stage for every side nozzle, such as to and from an intercooler, side gas injection, etc. For many applications the compression ratio across a frame will run 2.5–4.0.

One word of caution: For gas mixtures or air, be sure to include the contained water. It can be a healthy percentage for a Gulf Coast air compressor.

STEAM TURBINES

The process engineer can determine the steam requirements for turbines by knowing the horsepower. The theoretical steam rate can be determined from Reference 5 or Mollier charts following a constant entropy path. The theoretical steam rate in Reference 5 is given as lb/hr/kw, which is easily converted to lb/hr/hp. One word of caution: In using Reference 5, steam pressures are given in psig. Sea level is the basis. For low steam pressures at high altitudes, appropriate corrections must be made. See atmospheric pressure versus altitude in Chapter 7.

The theoretical steam rate (sometimes referred to as the "water rate") must then be divided by an adiabatic efficiency to obtain the actual steam rate. In lieu of manufacturer's data, use the following for preliminary work:

Horsepower	Efficiency, %
500–1,000	50
1,000–1,500	55
1,500–2,000	60
2,000–3,000	65
3,000–5,000	70
5,000 UP	75

The smaller turbines can vary widely in efficiency depending greatly on speed, horsepower, and pressure conditions. Very rough efficiencies to use for initial planning below 500 hp at 3,500 rpm are

Horsepower	Efficiency, %
1–10	15
10–50	20
50–300	25
300–350	30
350–500	40

Some designers limit the speed of the cheaper small steam turbines to 3600 rpm.

REFERENCES

1. *Elliott Multistage Compressors,* Bulletin P-25, Carrier Corporation. 1973.
2. *NGPSA Engineering Data Book,* 9th ed., Natural Gas Processors Suppliers Association, 1972.
3. Perry and Chilton, *Chemical Engineer's Handbook,* 5th ed., McGraw-Hill Inc., 1973.
4. Ludwig, Ernest, *Applied Process Design for Chemical and Petro-Chemical Plants,* Vol. 3, 3rd ed., Gulf Publishing Company, Houston, Tex., 2000 (in production).
5. Keenan, J. H. and F. G. Keyes, *Theoretical Steam Rate Tables, Trans. A.S.M.E.,* 1938.

4
Vacuum Systems

This chapter primarily deals with vacuum jets (eductors) and vacuum systems associated with vacuum fractionators. The design of vacuum overhead condensers is discussed in Chapter 7.

VACUUM JETS

Jet design is normally handled by the vendor. However, the process engineer must specify the system into which the jets are incorporated. He must also supply the vendor with operating conditions which include

1. Flows of all components to be purged from the system (often air plus water vapor).
2. Temperature and pressure entering the jets and pressure leaving if not atmospheric.
3. Temperature and pressure of steam available to drive the jets.

4. Temperature and quantity of cooling water available for the intercondensers. Also cooling water allowable pressure drop for the intercondensers.

In addition, the process engineer must be aware of good design practices for vacuum jets.

The vendor will convert the component flow data into an "air equivalent." Because jets are rated on air handling ability, he can then build up a system from his standard hardware. The vendor should provide air equivalent capability data with the equipment he supplies. Air equivalent can be determined with Equation 4-1.

$$ER = F\sqrt{0.0345\,(MW)} \tag{4-1}$$

where ER = entrainment ratio (or air equivalent). It is the ratio of the weight of gas handled to the weight of air that would be handled by the same ejector operating under the same conditions.

MW = gas mol wt

F = 1.00, for MW 1–30

F = 1.076 – 0.0026 (MW), for MW 31–140

Equation 4-1 will give results within 2% of the Reference 4 entrainment ratio curve.

Equations 4-2 and 4-3 show the effect of temperature.

$$ERTA = 1.017 - 0.00024T \tag{4-2}$$

$$ERTS = 1.023 - 0.00033T \tag{4-3}$$

where ERTA = the ratio of the weight of air at 70°F to the weight of air at a higher temperature that would be handled by the same ejector operating under the same conditions.

ERTS = same as above for steam

T = gas temperature, °F

This information is based on Reference 4.

The vendor should also supply steam consumption data. However, for initial planning the process engineering must have an estimate. Use Equation 3-4 or 3-5 to calculate the horsepower required to compress non-condensing components from the jet inlet pressure and temperature to the outlet pressure. For process water vapor handled by the jets with intercondensing, calculate horsepower for the first stage only. After the first stage, the condenser will bring the system to the same equilibrium as would have occurred without the process water vapor. Use an adiabatic efficiency of 7% for cases with jet intercondensers and 4% for non-condensing cases. Estimate the steam consumption to be the theoretical amount that can deliver the previously calculated total horsepower using the jet system steam inlet and outlet conditions. These ballpark results can be used until vendor data arrive. This procedure will give conservative results for cases with high water vapor compared to the Ludwig[2] curves for steam consumption.

Following are some general rules of thumb for jets:

1. To determine number of stages required, assume 7:1 compression ratio maximum per stage.
2. The supply steam conditions should not be allowed to vary greatly. Pressure below design can lower capacity. Pressure above design usually doesn't increase capacity and can even lower capacity.
3. Use Stellite or other hard surface material in the jet nozzle. For example, 316 s/s is insufficient.
4. Always provide a suitable knockout pot ahead of the jets. Water droplets can quickly damage a jet. The steam should enter the pot tangentially. Any condensate leaves through a steam trap at the bottom. It is a good idea to provide a donut baffle near the top to knock back any water creeping up the vessel walls.
5. The jet barometric legs should go in a straight line to the seal tank. A 60–90° slope from horizontal is best.

DESIGN RECOMMENDATIONS

First estimate air or other gas leakage into the system. Of course, every effort should be made to keep it as tight as possible. Possible leak points can be sealed with polystyrene, which produces an excellent seal. When tests cannot be made, use rules of thumb. Many such rough estimating techniques exist.

A close friend with years of experience in designing and operating vacuum systems claims that a system can be made very tight when designed properly. He advocates purging instrument leads with nitrogen. The rate is 1 scfm maximum per lead. He has designed and successfully operated a number of vacuum systems with provisions for the required instrument purge times a safety factor of about 4. This factor is a contingency for such things as sudden load changes that could bring in fractionator feed that has been more poorly stripped.

Other recommendations are

1. For good control, design the pressure drop for the control valve between the fractionating system and the jet system for sonic velocity (approximately 2:1 pressure ratio). This means that the jets' suction must be designed for half the absolute pressure of the evacuated system.
2. In even a large fractionator system, under the following conditions:
   - properly designed condenser (see Chapter 7)
   - tight vacuum
   - sonic velocity across the control valve
   - well stripped feed

   the control valve trim rarely exceeds ¾ in. and usually runs ½ in.
3. Set the pressure controller for low proportional band.

4. For applications having column temperature control above the feed point, put the measuring elements for the temperature and pressure controllers on the same tray. This will make for good composition control at varying column loads (varying column differential pressure). In vacuum systems, a slight pressure change will produce large equilibrium temperature changes.
5. For big vapor lines and condensers (frequent in vacuum systems), always insulate the line, condenser, and top of the column. Otherwise, rain or sudden cold fronts will change column control. It is possible to have more surface in the overhead line than in the condenser.
6. Never use screwed fittings in any vacuum system regardless of size.
7. Installation of a pressure controller measurement tap is shown in Figure 4-1.
8. Avoid liquid traps in vacuum system piping by never going up after having gone horizontal.
9. Put the vacuum system control valves at the highest point of a horizontal run and the control valve bypass in the same horizontal plane. This is in compliance with Item 8.

REFERENCES

1. *Cameron Hydraulic Data,* 13th ed., Ingersoll-Rand Company, Cameron Pump Division, 1965.

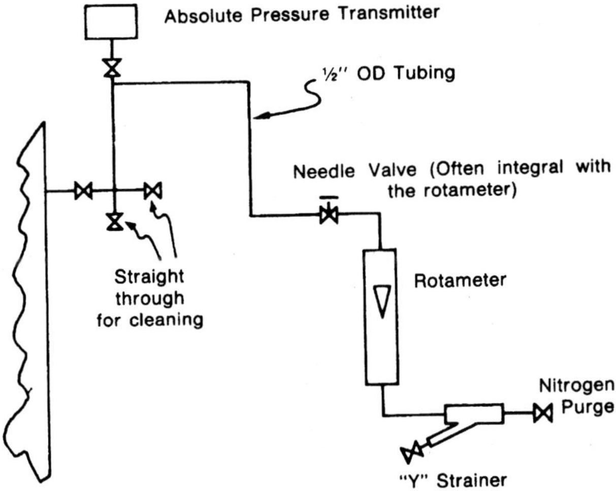

Figure 4-1. Vacuum measurement installation.

2. Ludwig, Ernest, *Applied Process Design for Chemical and Petro-Chemical Plants,* Vol. 1, 3rd ed. Gulf Publishing Company, Houston, Tex., 1995.
3. Perry and Chilton, *Chemical Engineers' Handbook,* 5th ed.
4. *Standards for Steam Jet Ejectors,* 3rd ed., Heat Exchange Institute, New York, N.Y., 1956.
5. Jackson, D. H., "Selection and Use of Ejectors," *Chemical Engineering Progress,* 44, 1948, 347.

5
Flow Component Sizing

Figure 5-1 illustrates a method that will produce a system in which the parts fit together to accomplish the common goal of good control. Control valve share of total system flowing pressure drop will be 60% at normal flow. The system will still achieve maximum flow as long as the control valve trim selected can pass maximum flow at operating head loss (line 23 of Figure 5-1). The procedure described in Figure 5-1 is intended as a stand-alone device for guiding the calculations, and worksheets can be prepared from it.

	LINE NO.				
1	Material				
2	Phase L, G, V, S				
3	Max. Flow 1000 lb/hr				
4	Norm. Flow 1000 lb/hr				
5	Min. Flow 1000 lb/hr				
6	Specific Gravity (base)				
7	Specific Gravity (flow)				
8	Mol. Wt.				
9	Temp. Flowing °F				
10	Temp. Base °F				
11	GPM Actual Fluid	Based on max flow.			
12	Press. Nominal (psia)				
13	P_2 Destination (psia)	Press is measured at liquid level.			
14	P_1 Source (psia)				
15	ΔH_p = [(13)–(14)] [N]/(7) ft	N = 2.307 ft. H_2O/psi			
16	Z_2 Disch. ft	Height of liquid level using a common base line.			
17	Z_1 Suct. ft	Use highest liquid height whether in pipe or vessel.			
18	ΔH_Z = (16)–(17) ft				
19	ΔH_1 Equip. ft	ΔH_1 = PSI × 2.307 / (7)		(Based on (11))	
20	ΔH_2 Line (Disch) ft				
21	ΔH_3 Line (suct) ft	Based on (11)			
22	ΔH_F (19) + (20) + (21) ft				
23	ΔH_{CV} = (x) × (22) ft	Value of x = 1.5 ((4)/(3))^2			
24	ΔH_{total} = (22) + (23) ft	Defined as total flowing loss.			
25	Pump TDH = (15) + (18) + (24) + meter permanent loss, ft				
26	Pump HP (Hydraulic)	HP = (3) × (25) × 5.05 × 10^{-4}			
27	Actual Pump HP = (26) / efficiency				
28	Suct Line IPS/SCH.				
29	Disch Line IPS/SCH.				
30	Suct Line Length, ft				
31	Disch Line Length, ft				
32	Meter Range, "H_2O				
33	d/D Orifice				
34	Full Scale Rdg 1000 lb/hr				

ΔH_p = Static pressure difference in ft. of fluid at flowing conditions.

ΔH_Z = Liquid head difference from source to destination in ft. of fluid at flowing conditions.

ΔH_1 = Flowing loss in exchangers, etc. in ft. of fluid at flowing conditions.

NOTE: All ft. are reported as ft. of flowing fluid at S.G. corresponding to (9).

Figure 5-1. A consistent method for flow component sizing.

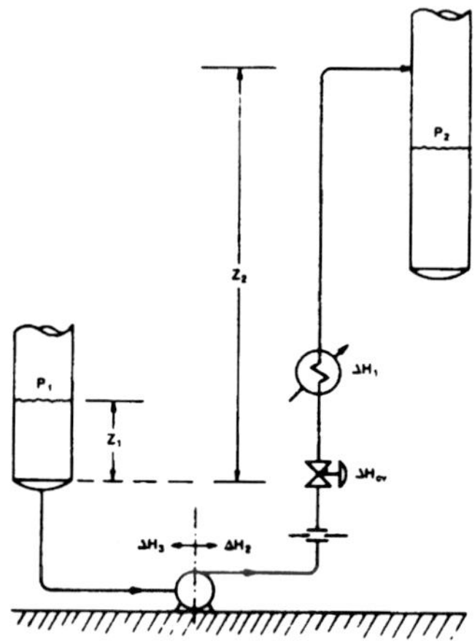

ΔH_2	= Flowing loss in discharge line in ft. of fluid at flowing conditions.
ΔH_3	= Flowing loss in suction line in ft. of fluid at flowing conditions.
ΔH_{CV}	= Pressure drop across control valve equal to 60% of total flowing loss at normal flow.
ΔH_F	= Total flowing loss excluding control valve and meter loss in ft. of fluid at flowing conditions.

Figure 5-1. *Continued.*

6
Pneumatic Conveying

TYPES OF SYSTEMS

This chapter considers conveying solids with air. Materials conveyed with air vary greatly in properties, often requiring special considerations. Therefore, this chapter gives only the means for preliminary rough sizing and selection of equipment. Often the entire pneumatic conveying system for a plant is bought as a package from a vendor specializing in such equipment. Even then it is helpful for the process engineer to be familiar with rough equipment sizing methods. Such knowledge can help him plan such things as plot area and utility draws in the initial design phase. The knowledge may also be helpful in later startup or troubleshooting. The methods described are primarily those of References 1–3.

The following are the general types of systems and their uses:

1. Negative (vacuum) system. Normally used when conveying from several pickup points to one discharge point.
2. Positive-pressure system. Normally used when conveying from one pickup point to several discharge points.
3. Pressure-negative (push-pull) combination system. Normally used when conveying from several pickup points to several discharge points.
4. Venturi, product systems, or blow tanks will not be discussed here.

The final choice is determined by economics or some special material demand.

Negative System

The negative system usually sucks on a cyclone and/or filter/receiver mounted above the receiving storage hopper or bin. Solids are usually sent down to the hopper with a rotary air lock feeder. Air is sucked into the transfer pipe or duct at the pickup end of the system. A variety of feeders can be used to introduce solids into the flowing air stream such as rotary air lock feeders, pan-type manifolds under rail car hoppers, paddle-type rail car unloaders, screw conveyors, etc.

Positive-Pressure System

In the positive-pressure system, air is blown into the pickup duct often up to a cyclone with atmospheric vent.

Usually solids are introduced to the conveying air stream with a rotary air lock feeder. In sending the solids down to the storage bin from the cyclone, a simple spout connection can be used instead of the rotary airlock feeder required in the negative-pressure system. The positive system doesn't need a vacuum vessel at each receiving location. Also in the positive-pressure system conveying to a number of hoppers, a simple bag-type cloth can serve as the filter. So for conveying from one pickup location to several receiving locations, the pressure system is often cheaper than the negative system.

Pressure-Negative System

The pressure-negative system is ideal for unloading rail cars and also for conveying light dusty materials because the suction at the inlet helps the product enter the conveying line. Positive systems are poorer than negative for handling such materials because feeding into the pressure line can be difficult and can present a dust-collection problem at the pickup location because of blowback or leakage air through the rotary valve. Also the negative system works better if there are lumps at the pickup end. The positive pressure system tends to pack at the lump, while the negative system will often keep solids moving around the lump and gradually wear it away.

For the pressure-negative system, a single blower can be used for both the negative and positive sides of the system. However, the lack of flexibility with a single blower usually dictates the need for separate blowers for the negative and positive sides.

DIFFERENTIAL PRESSURES

If the system differential pressure requirements are low, a fan can often be used. A fan is limited to a maximum of about 65 in. of water (just over 2 psi) in vacuum service or 77 in. of water in pressure service (just under 3 psi). To estimate fan horsepower use Equation 3-4 or 3-5. In lieu of manufacturer's data, use an adiabatic efficiency of 50% for initial work.

For higher differential pressures (or lower differential pressures when a fan is not preferred), the rotary positive-displacement blower is used. These are excellent for conveying systems because they provide:

1. reasonably constant volume at variable pressure discharge.
2. vacuums of about 8 psi (some can go to 11 psi with water injection) and pressure differentials of 15 psi.
3. a low slippage with improved efficiency.

The higher pressure differentials allow longer and sometimes smaller lines than for fans. To estimate blower horsepower, use Equation 3-4 or 3-5. In lieu of manufacturer's data, use an adiabatic efficiency of 80% for initial work.

Table 6-1.
Capacity Range

Duct Dia. in.	Flow, scfm at 5,000 ft/min	Friction loss, in H_2O/100 ft	Usual Capacity, thousands of lb/hr Negative	Positive
4	440	11.0	2–6	12–40
5	680	8.0	3–10	15–60
6	980	6.3	4–15	20–80
8	1,800	4.5	15–30	30–160

Even though 15 psi is possible from a blower, most positive-pressure systems are limited to 10–12 psi differential pressure because of limits of the rotary air lock feeder valves (deflections in shaft and bearings and increased blowback air).

EQUIPMENT SIZING

To rough out line sizes and pressure drop for fan or blower sizing, use the following quick method:

1. Arbitrarily assume air velocity of 5,000 ft/min (good for 90% of conveying situations).
2. Use Table 6-1.
3. Calculate pressure drop. This is in two parts:
   a. Material losses

      E_1 = acceleration losses

      E_2 = lifting energy

E_3 = horizontal losses
E_4 = bends and elbows

b. Air losses

4. First, calculate material losses in ft-lb/min:

$E_1 = MU^2/2g = 108M@5{,}000$ ft/min
$E_2 = M(H)$
$E_3 = M(L)(F)$
$E_4 = MU^2/gR(L)(F)(N) = 342(M)(F)(N)$ for 48-in. radius 90° ell. Assume this to be the case.

where M = solids conveyed, lb/min
U = velocity, ft/min
2g = 2.32×10^5 ft/min^2
H = vertical lift, ft
L = duct horizontal length, ft
R = 90° ell radius, ft
F = coefficient of friction and tangent of solids "angle of slide" or "angle of repose." Use 0.8 in lieu of solids data for initial estimating.
N = number of 90° ells. For 45°, 30°, etc., express as equivalent 90° ells by direct ratio (example: A 30° ell is 0.33 of a 90° ell).

5. From Table 6-1 and solids rate, estimate duct size and flow in scfm (standard cubic feet per minute).
6. Express material losses in inches of water:

$$\frac{\text{ft - lb / min}}{\text{ft}^3\text{ / min} \times 5.2} = \text{in. } H_2O$$

7. Calculate air losses.
   a. Calculate equivalent length of straight pipe by adding to the actual length of straight pipe an allowance for conveying type 90° ells of 1 ft of pipe/in. of diameter (example: 4-in. 90° ell = 4 ft of pipe).
   b. Assume the following losses for other items in inches of water:

duct entry loss	1.9
Y branch	0.3
cyclone	3.0
collector vessel	3.0
filter	6.0

8. Add material and air losses.
9. Calculate fan or blower horsepower as explained earlier.
10. Be sure to use the atmospheric suction pressure at the site. Normal blower ratings are given at sea level.

REFERENCES

1. Fisher, John, "Practical Pneumatic Conveyor Design," *Chemical Engineering,* McGraw-Hill, Inc., June 2, 1958. Used with permission.
2. Gerchow, Frank J., "How to Select a Pneumatic Conveying System," *Chemical Engineering,* McGraw-Hill, Inc., Feb. 17, 1975. Used with permission.
3. Perkins, Don E. and Jim E. Wood, "Design and Select Pneumatic Conveying Systems," *Hydrocarbon Processing,* March 1974.

7

Heat Transfer

TUBESIDE PRESSURE DROP

This pressure drop is composed of several parts that are calculated as shown in Tables 7-1 and 7-2.[1]

SHELLSIDE PRESSURE DROP

Tube Patterns

With segmental baffles, where the shellside fluid flows across the tube bundle between baffles, the following tube patterns are usual:

1. Triangular. Joining the centers of 3 adjacent tubes forms an equilateral triangle. Any side of this triangle is the tube pitch c.
2. Square inline. Shellside fluid has straight lanes between tube layers, unlike triangular where alternate tube layers are offset. This pattern makes for easy

Table 7-1
Calculation of Tubeside Pressure Drop in Shell and Tube Exchangers

Part	Pressure Drop in Number of Velocity Heads	Equation
Entering plus exiting the exchanger	1.6	$\Delta h = 1.6 \frac{U_P^{\,2}}{2g}$ (This term is small and often neglected)
Entering plus exiting the tubes	1.5	$\Delta h = 1.5 \frac{U_T^{\,2}}{2g} N$
End losses in tubeside bonnets and channels	1.0	$\Delta h = 1.0 \frac{U_T^{\,2}}{2g} N$
Straight tube loss	See Chapter 1 "Piping Pressure Drop"	

Δh = *Head loss in feet of flowing fluid*
U_p = *Velocity in the pipe leading to and from the exchanger, ft/sec*
U_T = *Velocity in the tubes*
N = *Number of tube passes*

cleaning because a lance can be run completely through the bundle without interference. This pattern has less pressure drop than triangular, but shell requirements are larger and there is a lower heat trans-

Table 7-2
Calculation of Tubeside Pressure Drop in Air-Cooled Exchangers

Part	Pressure Drop in Approximate Number of Velocity Heads	Equation
All losses except for straight tube	2.9 (calculated from curve on p. 9–11 of Reference 3)	$\Delta h = 2.9 \frac{U_T^2}{2g} N$
Straight tube loss	See Chapter 1 "Piping Pressure Drop"	

fer coefficient for a given velocity at many velocity levels. Joining the centers of 4 adjacent tubes forms a square. Any side of this square is the tube pitch c.

3. Square staggered, often referred to as square rotated. Rotating the square inline pitch 45° no longer gives the shellside fluid clear lanes through the bundle. Tube pitch c is defined as for square inline.

Two other terms need definition: transverse pitch a and longitudinal pitch b. For a drawing of these dimensions see Reference 1. For our purposes, appropriate lengths are shown in Table 7-3.

Table 7-3
Tube Pattern Relationships

	Triangular	Square Inline	Square Staggered
a	a = c	a = c	a = 1.414c
b	b = 0.866c	b = c	b = 0.707c

Turbulent Flow

For turbulent flow across tube banks, a modified Fanning equation and modified Reynolds number should be used. The following method is based on Reference 1.

$$\Delta P_f = \frac{4f''N_R N_{sp}\rho U_{max}^2}{2g} \tag{7-1}$$

$$Re' = \frac{D_o U_{max}\rho}{\mu} \tag{7-2}$$

where ΔP_f = friction loss in lb/ft^2
f'' = modified friction factor
N_R = rows of tubes per shell pass (N_R is always equal to the number of minimum clearances through which the fluid flows in series. For square staggered pitch, the maximum velocity, U_{max}, which is required for evaluating Re′, may occur in the transverse clearances a or the diagonal clearances c. In

the latter case, N_R is one less than the number of tube rows.

N_{sp} = number of shell passes

ρ = density, lb/ft^3

U_{max} = maximum linear velocity (through minimum cross-sectional area), ft/sec

g = 32.2 ft/sec^2

Re′ = modified Reynolds number

D_o = outside tube diameter, ft

μ = viscosity, lb/ft sec; centipoises × 0.000672

The modified friction factor can be determined by using Tables 7-4 and 7-5.

Table 7-4
Determination of f″ for 5 Tube Rows or More

C/D_o Both in Same Length Units	Triangular			
$Re' \times 10^{-3}$	2	8	20	40
1.25 (min)	0.210	0.155	0.130	0.107
1.50	.145	.112	.090	.074
2.00	.118	.096	.081	.066
3.00	.089	.076	.063	.052
	Square Inline			
$Re' \times 10^{-3}$	2	8	20	40
1.25 (min)	0.139	0.135	0.116	0.099
1.50	.081	.079	.080	.071
2.00	.056	.057	.055	.053
3.00	.052	.050	.045	.038

Table 7-4 (Continued)
Determination of f″ for 5 Tube Rows or More

C/D_o Both in Same Length Units	Square Staggered			
$Re' \times 10^{-3}$	2	8	20	40
1.25 (min)	0.130	0.106	0.088	0.063
1.50	.125	.103	.079	.061
2.00	.108	.090	.071	.058
3.00	....	....	....	....

Table 7-5
f″ Correction Factor For Less Than 5 Tube Rows

Number of Rows	1	2	3	4
Correction Factor	1.30	1.30	1.15	1.07

Laminar Flow

Below

$\dfrac{D_c U_{max} \rho}{\mu} = 40$ where D_c is the tube clearance in feet,

the flow is laminar. For this region, use

$$\Delta P_f = \frac{1.68 \mu U_{max} L}{D_e^2} \tag{7-3}$$

where L = length of flow path, ft
D_e = equivalent diameter, ft; 4 times hydraulic radius

$$D_e = 4 \frac{\text{(cross-sectional flow area)}}{\text{(wetted perimeter)}} = D_o \left(\frac{4ab}{\pi D_o^2} - 1 \right) \tag{7-4}$$

Pressure Drop for Baffles

Equations 7-1 and 7-3 determine the pressure drop across the tube bundle. For the additional drop for flow through the free area above, below, or around the segmental baffles, use

$$\Delta P_f = \frac{W^2 N_B N_{sp}}{\rho S_B^2 g} \quad (7\text{-}5)$$

where W = flow in lb/sec

N_B = number of baffles in series per shell pass

S_B = cross-sectional area for flow around segmental baffle, ft^2

Flow Parallel to Tubes

For flow parallel to tubes or in an annular space, for example a double-pipe heat exchanger, use

$$\Delta P = \frac{2f\rho U^2 L}{gD_e} \quad (7\text{-}6)$$

where ΔP = pressure drop, lb/ft^2

f = Fanning friction factor (see Chapter 1)

AIR-COOLED EXCHANGERS—AIRSIDE PRESSURE DROP

This method will approximate required fan horsepower based on Reference 3.

$$\Delta P_a = \frac{F_p N}{D_R} \quad (7\text{-}7)$$

Table 7-6
Determination of F_p

Air Face Mass Velocity lb/hr/ft² Face Area	F_p
1,400	0.033
1,600	.0425
1,800	.0525
2,000	.0625
2,200	.075
2,400	.0875
2,600	.100
2,800	.115
3,000	.130
3,200	.150
3,400	.165
3,600	.185

where ΔP_a = static pressure drop, in. H_2O
F_p = static pressure drop factor from Reference 3 (see Table 7-6)
N = number of tube rows

$$D_R = \frac{\text{Actual density at average air temperature}}{\text{Air density at sea level and 70°F}}$$

Use perfect gas law to calculate D_R

$$\text{Altitude (above sea level), ft} = 25{,}000 \ln\left(\frac{14.7}{P}\right)$$

where P = atmospheric pressure in psia

Once ΔP_a is obtained, the pressure that the fan has to provide is then calculated.

$$P_F = \Delta P_a + \left[\frac{\text{acfm / fan}}{3{,}140\,D^2}\right]^2 D_R \quad (7\text{-}8)$$

where P_F = total pressure that fan has to provide, inches of water

acfm = actual cubic feet per minute

D = fan diameter, feet

$$D_R = \frac{\text{actual density at fan temperature}}{\text{air density at sea level and } 70°\text{F}}$$

$$HP = \frac{\text{acfm / fan}(P_F)}{4{,}460} \quad (7\text{-}9)$$

where HP = approximate horsepower per fan

DETERMINING SHELL AND TUBE HEAT EXCHANGER SHELL DIAMETER

For triangular pitch proceed as follows:

1. Draw the equilateral triangle connecting three adjacent tube centers. Any side of the triangle is the tube pitch (recall 1.25 D_O is minimum).
2. Triangle area is ½ bh, where b is the base and h is the height.
3. This area contains ½ tube.

4. Calculate area occupied by all the tubes.
5. Calculate shell diameter to contain this area.
6. Add one tube diameter all the way around (two tube diameters added to the diameter calculated above).
7. The resulting is minimum shell diameter. There is no firm standard for shell diameter increments. Use 2-in. increments for initial planning.

For square pitch proceed similarly.

DETERMINING HEAT EXCHANGER TEMPERATURE DIFFERENCE

Only countercurrent flow will be considered here. It is well known that the log mean temperature is the correct temperature difference to be used in the expression:

$$q = UA\Delta T_M \qquad (7\text{-}10)$$

where
- q = heat duty in Btu/hr
- U = overall heat transfer coefficient in Btu/hr ft^2°F
- A = tube surface area in ft^2
- ΔT_M = mean temperature difference in °F. For our case it is the log mean temperature difference.

$$\Delta T_M = \frac{\text{GTD} - \text{LTD}}{\ln(\text{GTD} / \text{LTD})} \qquad (7\text{-}11)$$

where
- GTD = greater temperature difference
- LTD = lesser temperature difference

When GTD/LTD < 2, the arithmetic mean is within about 2% of the log mean.

These refer to hot and cold fluid terminal temperatures, inlet of one fluid versus outlet of the other. For a heat exchanger with no phase change, the ΔT_M gives exact results for true countercurrent flow. Most heat exchangers, however, deviate from true countercurrent, so a correction factor, F, is needed.

These correction factors are given in various heat transfer texts and are based on Reference 8. In lieu of correction factor curves, use the following procedure to derive the factor:

1. Assume shellside temperature varies linearly with length.
2. For first trial on tubeside, assume equal heat is transferred in each pass with constant fluid heat capacity.
3. Using the end temperatures of each shell and tube pass, calculate ΔT_M for each tube pass. From this the fraction of total duty for each tube pass is determined.
4. For the new end temperatures, calculate the new ΔT_M for each tube pass.
5. The arithmetic average of the tube pass ΔT_M's is the ΔT_M corrected for number of passes. F = ΔT_M corrected/ΔT_M uncorrected.

This procedure will quickly give numbers very close to the curves.

One thing to be careful of in heat exchangers is a design having a so-called "temperature cross." An example is shown in Figure 7-1.

In Figure 7-1 the colder fluid being heated emerges hotter than the outlet temperature of the other fluid. For actual heat exchangers that deviate from true countercurrent flow, the following things can happen under temperature cross conditions:

1. The design can prove to be impossible in a single shell.
2. The correction factor can be quite low requiring an uneconomically large area.
3. The unit can prove to be unsatisfactory in the field if conditions change slightly.

For Figure 7-1, assuming one shell pass and two or more tube passes, the correction factor is roughly 0.7. This shows the undesirability of a temperature cross in a single shell pass.

The calculation procedure for temperature correction factors won't work for a temperature cross in a single shell pass, but this is an undesirable situation anyway.

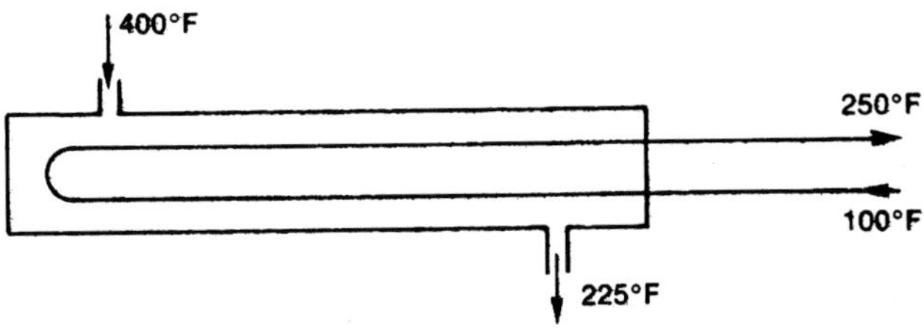

Figure 7-1. Example of a temperature cross.

Some conditions require breaking up the exchanger into multiple parts for the calculations rather than simply using corrected terminal temperatures. For such cases, always draw the q versus temperature plot to be sure no undesirable pinch points or even intermediate crossovers occur.

An example of a multisection calculation would be a propane condenser. The first section could be a desuperheating area, where q versus T would be a steeply sloped straight line followed by a condensing section with a straight line parallel to the q axis (condensing with no change in temperature). Finally, there could be a subcooling section with another sloped line. Calculate this unit as three separate heat exchangers.

DETERMINING HEAT EXCHANGER HEAT TRANSFER COEFFICIENT

To do this, sum all the resistances to heat transfer. The reciprocal of this sum is the heat transfer coefficient. For a heat exchanger, the resistances are

Tubeside fouling	R_{FT}
Shellside fouling	R_{FS}
Tube metal wall	R_{MW}
Tubeside film resistance	R_T
Shellside film resistance	R_S

For overall tubeside plus shellside fouling, use experience factors or 0.002 for most services and 0.004 for extremely fouling materials. Neglect metal wall resistance for overall heat transfer coefficient less than 200 or heat flux less than 20,000. These will suffice for ballpark work.

For film coefficients, many situations exist. For the purposes of this handbook, Table 7-7 gives ballpark estimates of film resistance at reasonable design velocities.

Tables 7-8 and 7-9 give overall ballpark U values for air-cooled heat exchangers.

For liquid boiling, the designer is limited by a maximum flux q/A. This handbook cannot treat this subject in detail. For most applications assuming a limiting flux of 10,000 will give a ballpark estimate.

(text continued on page 88)

Table 7-7
Film Resistances

Liquids	R
Water	0.0013
Gasoline	.0067
Gas oils	.0115
Viscous oils heating	.0210
Viscous oils cooling	.0333
Organic solvents	.0036
Gases	
Hydrocarbons	
Low pressure	.0364
High pressure	.0200
Air	
Low pressure	.0500
High pressure	.0250

(table continued on next page)

Table 7-7. Continued

Vapors Condensing	
Steam	
No air	0.0006
10% air by volume	.0010
20% air by volume	.0040
Gasoline	
Dry	.0067
With steam	.0044
Propanes, butanes, pentanes	
Pure	.0033
Mixed	.0067
Gas oils	
Dry	.0133
With steam	.0090
Organic solvents	.0030
Light oils	.0033
Heavy oils (vacuum)	.0285
Ammonia	.0133
Evaporation	
Water	.0007
Organic solvents	.0050
Ammonia	.0033
Light oils	.0044
Heavy oils	.0333

Table 7-8
Typical Overall Heat-Transfer Coefficients for Air Coolers[3]

Service		Fintube ½ in. by 9		Fintube ⅝ in. by 10	
Water and water solutions		U_b	U_x	U_b	U_x
Engine jacket water (r_f = .001)		110	– 7.5	130	– 6.1
Process water (r_f = .002)		95	– 6.5	110	– 5.2
50-50 Ethyl glycol-water (r_f = .001)		90	– 6.2	105	– 4.9
50-50 Ethyl glycol-water (r_f = .002)		80	– 5.5	95	– 4.4
Hydrocarbon liquid coolers	Viscosity C_p	U_b	U_x	U_b	U_x
	0.2	85	– 5.9	100	– 4.7
	0.5	75	– 5.2	90	– 4.2
	1.0	65	– 4.5	75	– 3.5
	2.5	45	– 3.1	55	– 2.6
	4.0	30	– 2.1	35	– 1.6
	6.0	20	– 1.4	25	– 1.2
	10.0	10	– 0.7	13	– 0.6
Hydrocarbon gas coolers	Pressure, psig	U_b	U_x	U_b	U_x
	50	30	– 2.1	35	– 1.6
	100	35	– 2.4	40	– 1.9
	300	45	– 3.1	55	– 2.6
	500	55	– 3.8	65	– 3.0
	750	65	– 4.5	75	– 3.5
	1000	75	– 5.2	90	– 4.2

Table 7.8. *Continued*

Air and flue-gas coolers (Use one-half of value given for hydrocarbon gas coolers)				
Steam condensers (Atmospheric pressure and above)				
	U_b	U_x	U_b	U_x
Pure steam (r_f = .0005)	125	– 8.6	145	– 6.8
Steam with non-condensibles	60	– 4.1	70	– 3.3
HC condensers				
Range, °F	U_b	U_x	U_b	U_x
0° range	85	– 5.9	100	– 4.7
10° range	80	– 5.5	95	– 4.4
25° range	75	– 5.2	90	– 4.2
60° range	65	– 4.5	75	– 3.5
100° and over range	60	– 4.1	70	– 3.3
Other condensers				
	U_b	U_x	U_b	U_x
Ammonia	110	– 7.6	130	– 6.1
Freon 12	65	– 4.5	75	– 3.5

Note: U_b is overall rate based on bare tube area and U_x is overall rate based on extended surface.

Table 7-9
Typical Transfer Coefficients for Air-Cooled Heat Exchangers[6]

Condensing Service	U Btu/hr, ft^2 °F
Amine reactivator	90–100
Ammonia	100–120
Freon 12	60–80
Heavy naphtha	60–70
Light gasoline	80
Light hydrocarbons	80–95
Light naphtha	70–80
Reactor effluent—platformers, rexformers, hydroformers	60–80
Steam (0–20 psig)	130–140
Still overhead—light naphthas, steam, and non-condensible gas	60–70
Gas cooling service	
Air or flue gas @50 psig (ΔP = 1 psi.)	10
Air or flue gas @100 psig (ΔP = 2 psi.)	20
Air or flue gas @100 psig (ΔP = 5 psi.)	30
Ammonia reactor stream	80–90
Hydrocarbon gases @15–50 psig (ΔP = 1 psi.)	30–40
Hydrocarbon gases @ 50–250 psig (ΔP = 3 psi.)	50–60
Hydrocarbon gases @250–1,500 psig (ΔP = 5 psi.)	70–90

(table continued on next page)

Table 7-9 *Continued*

Liquid Cooling Service	U Btu/hr, ft² °F
Engine jacket water	120–130
Fuel oil	20–30
Hydroformer and platformer liquids	70
Light gas oil	60–70
Light hydrocarbons	75–95
Light naphtha	70
Process water	105–120
Residuum	10–20
Tar	5–10

Coefficients are based on outside bare tube surface for 1-in. O.D. tubes with eight extruded Al fins/in., ⅝ in. high, 16.9 surface ratio.

(text continued from page 83)

AIR-COOLED HEAT EXCHANGER ROUGH RATING

The preceding information will suffice to rough rate shell and tube units. For air-cooled heat exchangers, more is involved in rough rating. A suggested procedure is as follows:[2, 3]

1. Calculate exchanger duty (MMBtu/hr).
2. Select an overall U_x from Tables 7-8 and 7-9 (based on finned area). Arbitrarily use ½-in. fins, nine to the inch for determining U_x.

3. Calculate approximate air temperature rise from

$$\Delta T_a = \left(\frac{U_x + 1}{10}\right)\left(\frac{T_1 + T_2}{2} - t_1\right) \tag{7-12}$$

where ΔT_a = airside temperature rise, °F
U_x = overall coefficient based on finned area, Btu/hr ft^{2}°F
T_1, T_2 = process side inlet and outlet temperatures, °F
t_1 = air inlet temperature, °F

4. Calculate ΔT_M and apply appropriate correction factor F.
5. Calculate exchanger extended area from

$$A_x = \frac{q}{U_x \Delta T_M} \tag{7-13}$$

where A_x = extended (finned) surface, ft_2
q = duty, Btu/hr
ΔT_M = log mean temperature difference,°F

6. Estimate number of tube rows from Hudson Company optimum bundle depth curve, Figure 7-2. Use 4 to 6 tube rows if curve comes close to that number.
7. Arbitrarily choose 1-in. OD tubes, ½-in fins, 9 to the inch at 2-in. Δ pitch. This will give "middle of the road" face area. Use Table 7-10.

Table 7-10
Design Values for Rough Rating

Typical Section Width, ft	Typical Tube Length, ft
6	6, 10, 15, 20, 24, 30
8	10, 15, 20, 24, 30
12	12, 16, 24, 32, 40
16	16, 24, 32, 40

APF = extended area ft²/ft of tube = 3.80
AR = extended area/bare area = 14.5
APSF = extended area/bundle face area = No. tube rows (22.8)

8. Calculate face area (F_a), ft^2

$$F_a = \frac{A_x}{APSF}$$

9. Pick a desirable combination of tube length and section width to achieve the approximate face area.
10. Estimate number of fans. Use the fact that fanned section length divided by bay width seldom exceeds 1.8. A 16-ft wide bay with 24-ft tubes would have one fan (ratio = 1.5). The same 16-ft wide bay with 32-ft tubes would have 2 fans (ratio would be 2.0 for 1 fan).
11. Estimate minimum fan area by

$$FAPF = \text{fan area per fan} = \frac{0.40(F_a)}{\text{number of fans}}$$

12. From above, calculate fan diameter rounded up to the next even foot.

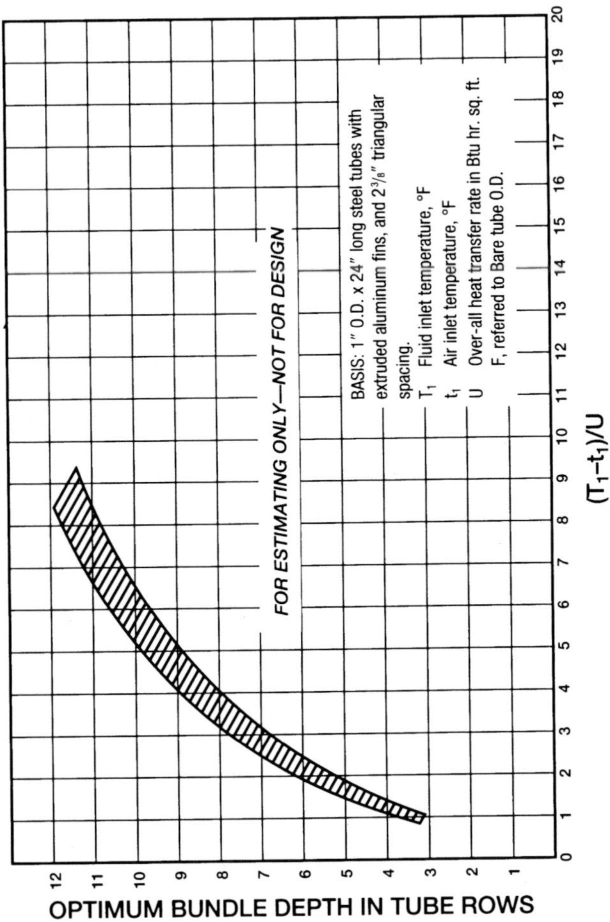

Figure 7-2. This is the effect of temperature level and overall transfer rate upon optimum bundle depth.[6]

13. Calculate horsepower per fan from the section in this chapter, "Air-Cooled Exchangers—Air Side Pressure Drop."
14. Number of tubes (N_T) can be obtained from

$$N_T = \frac{A_x}{APF\left(\text{Tube length}\right)}$$

VACUUM CONDENSERS

Outlet Temperature and Pressure. It is important to have proper subcooling in the vent end of the unit to prevent large amounts of process vapors from going to the vacuum system along with the inerts.

Control. It is necessary to have some over-surface and to have a proper baffling to allow for pressure control during process swings, variable leakage of inerts, etc. One designer adds 50% to the calculated length for the over-surface. The condenser must be considered part of the control system (similar to extra trays in a fractionator to allow for process swings not controlled by conventional instrumentation).

The inerts will "blanket" a portion of the tubes. The blanketed portion has very poor heat transfer. The column pressure is controlled by varying the percentage of the tube surface blanketed. When the desired pressure is exceeded, the vacuum system will suck out more inerts, and lower the percentage of surface blanketed. This will increase cooling and bring the pressure back down to the desired level. The reverse happens if the pressure falls below that desired. This is simply a matter of adjusting the heat transfer coefficient to heat balance the system.

Figure 7-3 shows typical baffling. The inerts move through the first part of the condenser, as directed by the baffles. The inerts then "pile-up" at the outlet end, lowering heat transfer as required by the controller. A relatively large section must be covered by more or less stagnant inerts that are sub-cooled before being pulled out as needed. Without proper baffles, the inerts build up in the condensing section and decrease heat transfer until the pressure gets too high. Then the vacuum valve opens wider, pulling process vapor and inerts into the vacuum system. Under these conditions, pressure control will be very poor.

Pressure Drop. Baffling must be designed to keep the pressure drop as low as possible. The higher the pressure drop, the higher the energy consumption and the harder the job of attaining proper vent end sub-cooling. Pressure drop is lower at the outlet end because of smaller mass flow.

By-Passing. Baffles should prevent by-pass of inlet vapor into the vent. This is very important.

Typical Condenser. Figure 7-3 illustrates an inlet "bathtub" used for low vacuums to limit pressure drop at the entrance to the exchanger and across the first rows of tubes. Note staggered baffle spacing with large spacing at inlet, and the side-to-side (40% cut) baffles. Enough baffles must be used in the inlet end for minimum tube support. In the last 25% of the outlet end, a spacing of 1⁄10 of a diameter is recommended.

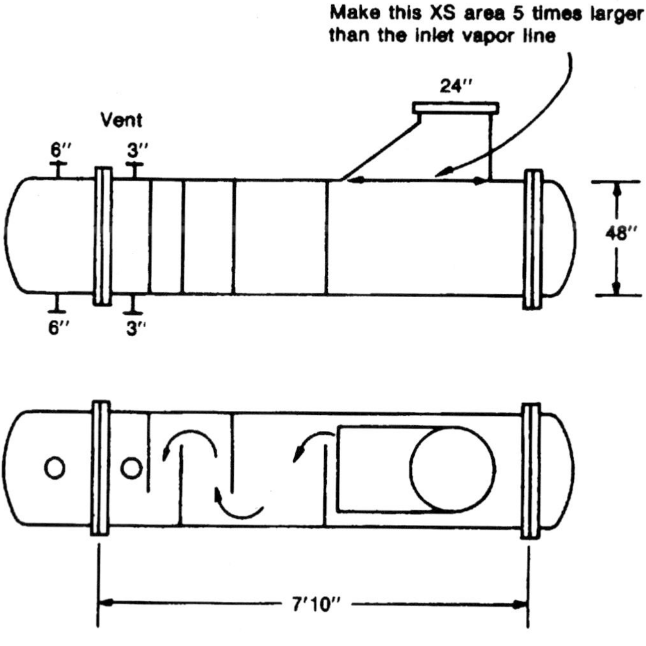

Figure 7-3. Baffling and inlet "bathtub" are shown in this typical vacuum condenser design.

FOGGING

Cause. Fogging occurs in a condenser when the mass transfer doesn't keep up with the heat transfer. The design must provide sufficient time for the mass transfer to occur. A high temperature differential (ΔT) with non-condensibles present or a wide range of molecular weights can produce a fog. The high ΔT gives a high driving force for heat transfer. The driving force for mass transfer, however, is limited to the concentration driving force (ΔY) between the composition of the condensible component in the gas phase and the composition in equilibrium with the liquid at the tube wall temperature. The mass transfer driving force (ΔY) thus has a limit. The ΔT driving force can, under certain conditions, increase to the point where heat transfer completely outstrips mass transfer, which produces fogging.

Nature of a Fog. Fog, like smoke, is a colloid. Once a fog is formed, it is very difficult to knock down. It will go right through packed columns, mist eliminators, or other such devices. Special devices are required to overcome a fog, such as an electric precipitator with charged plates. This can overcome the zeta potential of the charged particles and make them coalesce.

A colloid fog will scatter a beam of light. This is called the "Tyndall Effect" and can be used as a troubleshooting tool.

Cures. Eliminate the source of fogging. Use a smaller ΔT and thus more surface for mass transfer. Try to minimize $\Delta T/\Delta Y$.

Calculations. To check a design for possible fogging, the following procedure correctly considers mass transfer and heat transfer as two separate processes. Expressions for these are

Heat Transfer:

$$Q = UA\Delta T_M \quad \text{or} \quad Q = h_i A\Delta T_i \tag{7-14}$$

where i refers to the condensing side only.

Mass Transfer:

$$W = KA\Delta Y = KA(Y - Y_i) \tag{7-15}$$

where h_i = condensing side film coefficient, Btu/hr ft^2 °F
W = condensate rate, lb/hr
K = mass transfer coefficient, lb/hr ft^2 ΔY
Y = composition of the condensible component in the gas phase
Y_i = composition of the condensible component in equilibrium with liquid at tube wall temperature

Mass and heat transfer are related as follows:

$$\frac{h_i}{KC} = \left(\frac{K_T}{C\rho K_D}\right)^{2/3} \tag{7-16}$$

where C = heat capacity (gas phase), Btu/lb °F
ρ = gas density, lb/ft^3
K_T = thermal conductivity, Btu/hr ft^2 (°F/ft)
K_D = diffusivity, ft^2/hr

The ratio of heat transfer to mass transfer is:

$$\frac{Q}{W} = \frac{h_i \Delta T_i}{K \Delta Y} \tag{7-17}$$

because $\dfrac{h_i}{K} = C\left(\dfrac{K_T}{C\rho K_D}\right)^{2/3}$

then $\dfrac{Q}{W} = C\left(\dfrac{K_T}{C\rho K_D}\right)^{2/3} \dfrac{\Delta T_i}{\Delta Y}$

Note that Q refers only to sensible heat transfer. All latent heat is transferred via mass transfer. Likewise, h_i refers only to a dry gas coefficient (no condensation considered).

The calculations are made as follows. The exchanger is divided into small increments to allow numerical integrations. A tube wall temperature is first calculated and then Q/W. The gas temperature and composition from an increment can then be calculated. If the gas composition is above saturation for the temperature, any excess condensation can occur as a fog. This allows the degree of fogging tendency to be quantified. Whenever possible, experimental data should be used to determine the ratio of heat transfer to mass transfer coefficients. This can be done with a simple wet and dry bulb temperature measurement using the components involved.

HEAT EXCHANGER RULES OF THUMB

1. For fixed tubesheet design of shell and tube heat exchangers, don't allow too high a temperature difference between tubeside and shellside without providing a shellside expansion joint. The author has seen 70°F (one company) and 100°F (another company) used as this limit. An easy way to calculate the maximum stress is as follows:
   a. Assume the tubes are at tubeside bulk temperature and the shell is at shellside bulk temperature.
   b. Calculate the elongation of tubes, if unhampered, and shell, if unhampered, starting at 70°F and using the respective coefficients of expansion.
   c. If the tubes would have grown more or less than the shell, the difference will set up stress in both members, one in tension and the other in compression.
   d. Assume the deformation (strain) in each member is inversely proportional to its cross-sectional area. In other words, the fraction of the total strain in inches/inch of length will be proportionally more for the

member (tubes or shell) having the smallest cross section.

e. For each member, multiply its strain by Young's modulus (Modulus of Elasticity) in consistent units to get stress. Strain is dimensionless so Young's modulus in lb/in^2 will yield stress in lb/in^2.

f. Compare this with the maximum allowable stress for the material used.

g. The tensile and compressive moduli for steel are essentially the same.

2. Typical handling of design parameters.

DP = MOP + 10%, but not less than MOP + 30 psi

DP for vacuum, use 15 psi external pressure for cases having atmospheric pressure on the other side.

DT: Below 32°F DT = minimum operating temperature
Above 32°F DT = MOT + 25°F, but not less than 150°F

where DP = design pressure
DT = design temperature
MOP = maximum operating pressure
MOT = maximum operating temperature

3. 40% baffle cut = 40% open area is a good rule-of-thumb maximum for shell and tube heat exchangers.

REFERENCES

1. *Scovill Heat Exchanger Tube Manual,* Scovill Manufacturing Company, 1957.
2. Lerner, J. E., Fluor Corporation, "Simplified Air Cooler Estimating," *Hydrocarbon Processing,* Feb. 1972.
3. *NGPSA Engineering Data Book,* 9th ed., Natural Gas Processors Suppliers Association, 1972.
4. Ludwig, Ernest, *Applied Process Design for Chemical and Petrochemical Plants,* Vol. 3, 3rd ed., Gulf Publishing Company, Houston, Tex., 2000 (in production).
5. *Standards of Tubular Exchanger Manufacturers Association* (TEMA), 5th ed., 1968.
6. Smith, Ennis, C., Hudson Products Corporation, *Cooling with Air—Technical Data Relevant to Direct Use of Air for Process Cooling.*
7. Lee, D. C., J. W. Dorsey, G. Z. Moore, and F. D. Mayfield, "Design Data for Thermosiphon Reboilers," *Chemical Engineering Progress 52,* No. 4, 1956, p. 160.
8. Bowman, Mueller, and Nagle, *Trans. Am. Soc. Mech. Engrs.,* 62, 1940, p. 283.

8
Distillation

Because this subject is relatively complicated, only some quick estimating techniques will be discussed.

RELATIVE VOLATILITY

The equilibrium vaporization constant K is defined for a compound by

$$K_i = \frac{Y_i}{X_i} \tag{8-1}$$

where Y_i = mole fraction of component i in the vapor phase
X_i = mole fraction of component i in the liquid phase

To calculate a distillation, the relative volatility α is needed. It is defined as

$$\alpha = \frac{K_i}{K_j} \tag{8-2}$$

where i and j represent two components to be separated.

Raoult's Law for ideal systems is

$$p_i = P_i X_i \tag{8-3}$$

where p_i = partial pressure of i
P_i = vapor pressure of pure component i

By definition

$$p_i = \Pi Y_i \tag{8-4}$$

where Π = total pressure of the system

so $P_i X_i = \Pi Y_i$

and $\dfrac{P_i}{\Pi} = \dfrac{Y_i}{X_i} = K_i$

Therefore, for systems obeying Raoult's Law,

$$\alpha = \frac{P_i}{P_j} \tag{8 - 5}$$

Having a relative volatility, one can estimate theoretical distillation stages. There are a number of modern computer programs for distillation. These give excellent results, however, these must be applied by someone with a sound understanding of distillation in general and the accuracy of the input data in particular.

For our purposes, there are shortcut equations or it is fairly easy to produce a McCabe-Thiele diagram for binary systems. It helps to have a rough feel for the answer even if a computer program is to be run. For many splits, a McCabe-Thiele diagram is a real help in "seeing" how the system behaves.

MINIMUM STAGES

For binary systems or systems that approach binary, the Fenske-Underwood-Erbar/Maddox Method is recommended. For minimum stages, use the Fenske equation.[12]

$$N_m + 1 = \frac{\ln\left(\frac{X_{LD}}{X_{HD}}\right)\left(\frac{X_{HB}}{X_{LB}}\right)}{\ln \alpha_{average}} \qquad (8\text{-}6)$$

where N_m = minimum theoretical stages, not including reboiler (total reflux)

$X_{LD}, X_{HD}, X_{HB}, X_{LB}$ = mole fractions of light and heavy components in the distillate and bottoms

$\alpha_{average}$ = $\sqrt{\alpha_D \alpha_B}$, where differences are small in α_D and α_B. For values of α near 1.0, great care must be exercised because a small change in α has a large effect in the results.

MINIMUM REFLUX RATIO

For binary or near binary minimum reflux ratio, L/D min, use the Underwood equations.[10]

Bubble Point Liquid Feed

$$R_m = \frac{1}{\alpha - 1}\left[\frac{X_{LD}}{X_{LF}} - \frac{\alpha(1 - X_{LD})}{(1 - X_{LF})}\right] \tag{8-7}$$

Dew Point Vapor Feed

$$R_m = \frac{1}{\alpha - 1}\left[\frac{\alpha X_{LD}}{X_{LF}} - \frac{(1 - X_{LD})}{(1 - X_{LF})}\right] - 1 \tag{8-8}$$

where R_m = minimum reflux ratio
L = reflux rate, lb mols/hr
D = distillate rate, lb mols/hr
B = bottoms rate, lb mols/hr
F = feed rate, lb mols/hr

L, D, B, and F will be used later.

For a multicomponent system, a simple technique to yield conservative results is as follows. Combine a light key component and all lighter components, and a heavy key component and all heavier components into two groups to get X_F, X_D, X_B. For the key groups, use the α of the keys themselves.

A more accurate but longer method is to disregard the presence of components having a D/B greater than 100 or less than 0.01. Light and heavy key groups are generated. The group α's are found by plotting log D/B versus α with a straight line drawn through the major points. The α for each group is read at the D/B for the group. This procedure was introduced by Hengstebeck.[9]

ACTUAL REFLUX RATIO

For determining the theoretical stages at an actual reflux ratio, use the Erbar/Maddox relationship in Figure 8-1. N is the theoretical stages and R is the actual reflux ratio L/D. The method is defined in Reference 7.

The actual reflux ratio that is used should be optimized from economics data. For a ballpark estimate, use 1.1–1.2 Rm for a refrigerated system and 1.2–1.35 R_m for a hot system.

REFLUX-TO-FEED RATIO

Heretofore, the reflux ratio has been defined as reflux/distillate, L/D. Another very useful molar ratio is reflux/feed, L/F. In binary systems, L/F for all practical purposes is unchanging for wide differences in feed composition, so long as the following hold:

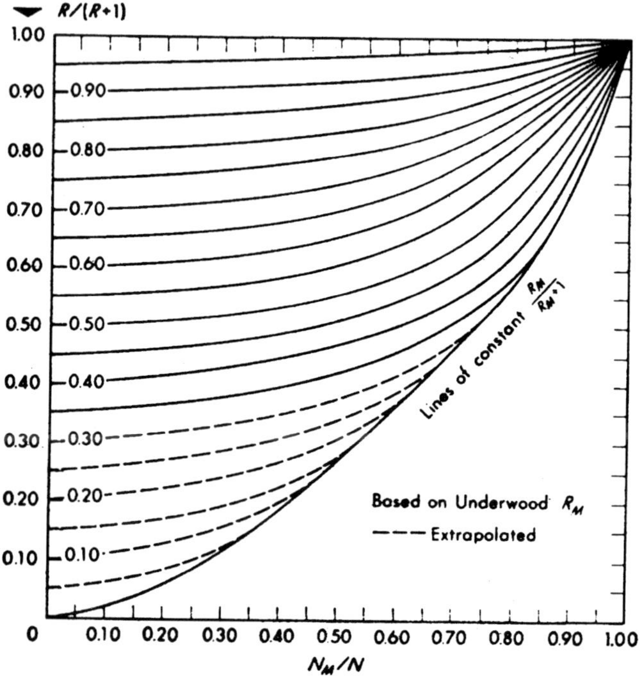

Figure 8-1. In this graph of the Erbar-Maddox relationship of reflux versus stages, N is the theoretical stages and R is the reflux ratio L/D.[7]

1. The distillate and bottoms compositions, but not necessarily the quantities, are held constant.
2. The feed tray is kept matched in composition to the feed (which means the feed tray moves with feed composition changes).

The reader can verify the above using the Underwood equations and the tower material balance. I once calculated a case where a large feed change would change L/D by 46%, whereas L/F changed only 1%. Several investigators report that the stability of L/F is well proven in the field. L/F is a good factor to use in predicting the effect of feed changes for design and in an operating plant.

MCCABE-THIELE DIAGRAM

In addition to the previously mentioned shortcut equations, plotting a McCabe-Thiele diagram is also a very useful tool. The equation for the equilibrium X-Y diagram and plotting of the operating lines are described next.

Equilibrium Curve

$$Y = \frac{X\alpha}{1 + (\alpha - 1)X} \qquad (8\text{-}9)$$

where Y = mole fraction of the light component in the vapor
X = mole fraction of the light component in the liquid

Another useful form is

$$\alpha = \frac{Y(1 - X)}{X(1 - Y)} \tag{8-10}$$

q Line—The Operating Lines Intersect on This Line

The thermal condition of the feed is designated as q, and is approximately the amount of heat required to vaporize one mole of feed at the feed tray conditions, divided by the latent heat of vaporization of the feed. One point on the q line is on the 45° line at X_F.

Bubble point liquid feed, q = 1.0
Dew point vapor feed, q = 0
General feed, $q = (L_S - L_R)/F$
Slope of q line = $q/(q - 1)$

where L_S = liquid rate in stripping section, lb mols/hr
L_R = liquid rate in rectifying section, lb mols/hr

Rectifying Section Operating Line

One point is on the 45° line at X_D, and the slope is L_R/V_R, where V_R is the rectifying section vapor rate. Another point is on the Y axis above the origin at DX_D/V_R.

Stripping Section Operating Line

One point is on the 45° line at X_B, and the slope is L_S/V_S, where S refers to the stripping section. Another point is the intersection of the rectifying section operating line with the q line. And still another point is on the Y axis at the location below the origin of $-(B/V_S)(X_B)$.

TRAY EFFICIENCY

The foregoing define the number of theoretical states. Actual stages depend upon the tray efficiency, which will probably be the weakest number in the design. Using operating data from a similar system is certainly best where possible. Table 8-1 gives some shortcut correlations.

Ludwig[1] discusses a work by the AIChE that has produced a method more detailed than the previous short-cut methods. He states that some of the shortcut methods can be off by 15–50% as indicated by the AIChE work. The spread of the Drickamer and Bradford correlation shown in the Ludwig plot is about 10 points of efficiency or ±5 efficiency points around the curve. Ludwig states that comparisons between shortcut and AIChE values indicate that deviations for the shortcut methods are usually on the safe or low side.

Maxwell's correlation was generated from hydrocarbon data only. Ludwig states that the Drickamer and Bradford correlation is good for hydrocarbons, chlorinated hydrocarbons, glycols, glycerine, and related compounds, and some rich hydrocarbon absorbers and strippers.

Table 8-1
Fractionator Overall Tray Efficiency, %

Viscosity Centipoises	Gunness[6] and Other Data Plotted Versus Reciprocal Viscosity in Maxwell[2] (Average viscosity of liquid on the plates)	Drickamer and Bradford[4] Correlation Plotted in Ludwig[1] (Molal average viscosity of the feed)
0.05*	...	98
0.10	104**	79
0.15	86	70
0.20	76	60
0.30	63	50
0.40	56	42
0.50	50	36
0.60	46	31
0.70	43	27
0.80	40	23
0.90	38	19
1.00	36	17
1.50	30	7
1.70	28	5

**Extrapolated slightly*
***Maxwell[2] explains how efficiencies above 100% are quite possible.*

Ludwig also presents correlations of O'Connell.[5] He warns that O'Connell may give high results. Ludwig suggests using the O'Connell absorber correlation only in areas where it gives a lower efficiency than the fractionator correlation of Drickamer and Bradford or O'Connell. The O'Connell correlations are given in Table 8-2.

For high values of α low tray efficiency usually results.

Table 8-2
Overall Tray Efficiency, % (O'Connell)

Correlating Factor	Fractionators*	Absorbers $\left(\frac{HP}{\mu} = \frac{\rho}{Km\mu}\right)^{**}$
0.01	. . .	8
0.05	. . .	17
0.10	87	22
0.15	80	23
0.20	74	26
0.30	67	29
0.40	62	32
0.50	57	33
0.60	55	35
0.70	52	36
0.80	51	37
0.90	49	38
1	48	39
1.5	44	42
2	41	45
3	37	48
4	35	52
5	33	53
6	32	56
7	32	57
8	31	58
9	. . .	60
10	. . .	61

**Fractionators = (relative volatility of key components) (viscosity of feed in centipoises). Relative volatility and viscosity are taken at average tower conditions between top and bottom.*

***H = Henry's law constant, mols/ft³ (atm)*

P = pressure, atm

μ = viscosity, centipoises

ρ = density, lbs/ft3

K = equilibrium K of key components

M = mol wt of liquid

F-FACTOR GUIDELINES FOR DISENGAGING

The F-factor is defined as $F = u\,(\rho)^{0.5}$,

where F = factor for allowable velocity
u = velocity, ft/sec
ρ = density, lb/ft^3

For quick estimates the following F-factor guidelines are proposed:

Situation	F-Factor
Fractionating column total cross section vapor velocity	1.0–1.5
Sieve tray hole velocity to avoid weeping	>12
Disengaging equipment for liquid/vapor separation	<6

DIAMETER OF FRACTIONATORS, GENERAL

For checking designs one can roughly relate tower diameter to reboiler duty as follows:

Situation	Reboiler Duty, MM/Btu/hr
Pressure distillation	$0.5\ D^2$
Atmospheric pressure distillation	$0.3\ D^2$
Vacuum distillation	$0.15\ D^2$

where D = tower diameter, ft

SMITH-BRINKLEY METHOD

The Smith-Brinkley Method uses two sets of separation factors for the top and bottom parts of the column, in contrast to a single relative volatility for the Underwood Method.[10] The Underwood method requires knowing the distillate and bottoms compositions to determine the required reflux. The Smith-Brinkley Method starts with the column parameters and calculates the product compositions. This is a great advantage in building a model for hand or small computer calculations. Starting with a base case, the Smith-Brinkley Method can be used to calculate the effect of parameter changes on the product compositions.

Smith[13] fully explains the Smith-Brinkley Method and presents a general equation from which a specialized equation for distillation, absorption, or extraction can be obtained. The method for distillation columns is discussed here.

For distillation component i,

$$f = \frac{\left(1 - S_n^{\,N-M}\right) + R\left(1 - S_n\right)}{\left(1 - S_n^{\,N-M}\right) + R\left(1 - S_n\right) + hS_n^{\,N-M}\left(1 - S_m^{\,M+1}\right)} \qquad (8\text{-}11)$$

where $f_i = (BX_B/FX_F)_i$

$S_{ni} = K_iV/L$

$S_{mi} = K'_iV'/L'$

h = Correlating factor; if the feed is mostly liquid, use Equation 8-12 and if mostly vapor, Equation 8-13

$$h_i = \frac{K'_i}{K_i}\left(\frac{L}{L'}\right)\left(\frac{1-S_n}{1-S_m}\right)_i \tag{8-12}$$

$$h_i = \frac{L}{L'}\left(\frac{1-S_n}{1-S_m}\right)_i \tag{8-13}$$

The effective top and bottom section temperatures are used to determine K_i and K'_i. These are used along with the effective top and bottom section molar liquid and vapor rates to determine S_n and S_m.

Nomenclature

- B = bottoms total molar rate or subscript for bottoms
- F = feed total molar rate or subscript for feed
- f_i = ratio of molar rate of component i in the bottoms to that in the feed
- K_i = equilibrium constant of component i in the top section equals y/x
- K'_i = equilibrium constant of component i in the bottom section equals y/x
- L = effective total molar liquid rate in top section
- L' = effective total molar liquid rate in bottom section
- M = total equilibrium stages below the feed stage including reboiler
- N = total equilibrium stages in the column including reboiler and partial condenser
- R = actual reflux ratio
- S_m = stripping factor for component i in bottom section

S_n = stripping factor for component i in top section
V = effective total molar vapor rate in top section
V′ = effective total molar vapor rate in bottom section
X = mol fraction in the liquid
Y = mol fraction in the vapor

REBOILERS

Reboiler Types

The *GPSA Engineering Data Book*[14] has an excellent section on reboilers. The most common types are the following:

- Forced circulation (Figure 8-2)
- Natural circulation (Figure 8-3)
  - Once-through
  - Recirculating
- Vertical thermosyphon (Figure 8-4)
- Horizontal thermosyphon (Figure 8-5)
- Flooded bundle (Kettle) (Figure 8-6)
- Recirculating—Baffled bottom (Figure 8-7)

For thermosyphon reboilers, the hydraulic aspects are as important as the heat transfer aspects. The design of thermosyphon reboiler piping is too broad a subject for this handbook. Some good articles on the subject can be found in References 15–27. Reference 16 is particularly good for horizontal thermosyphon reboilers. Table 8-3 has typical vertical thermosyphon design standards.

(text continued on page 120)

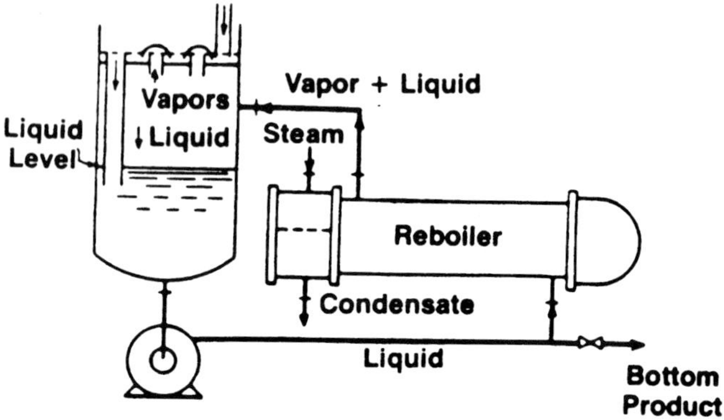

Figure 8-2. Forced-circulation reboiler arrangement.

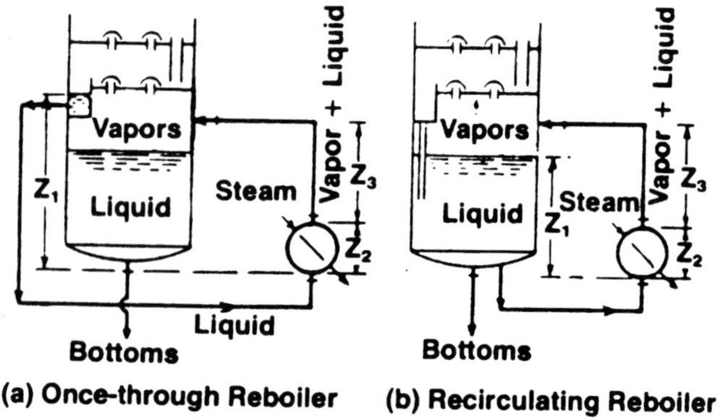

Figure 8-3. Natural-circulation reboiler arrangements.

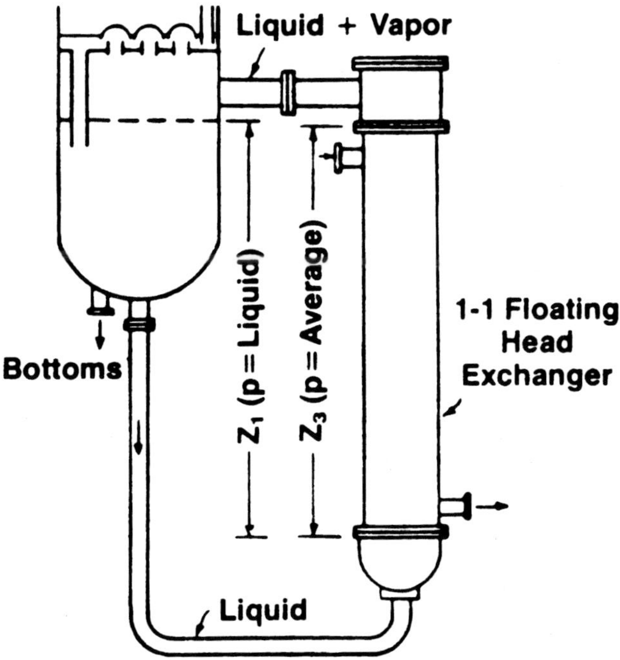

Figure 8-4. Vertical thermosyphon reboiler connected to tower.

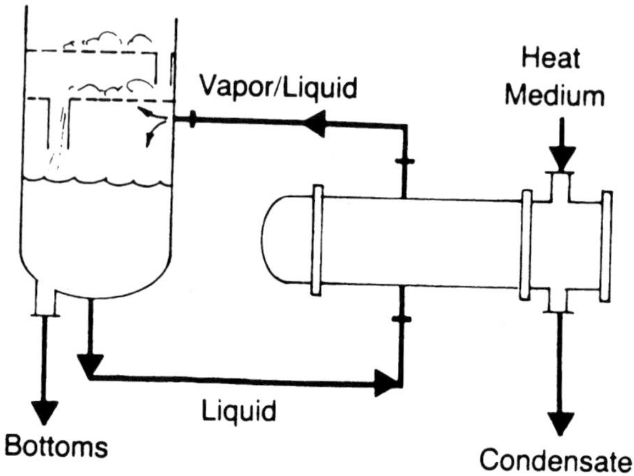

Figure 8-5. Horizontal thermosyphon reboiler.

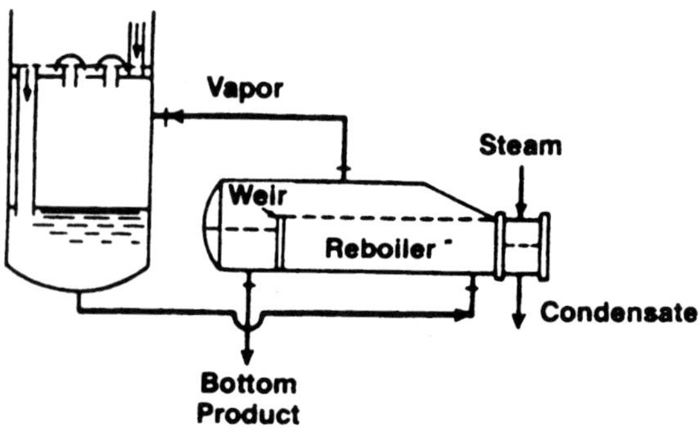

Figure 8-6. Kettle reboiler arrangement.

Table 8-3
Typical Thermosiphon Reboiler Design Standards*

Shell O.D., in.	No. tubes	Tube sheet faces ft-in.	Approx. ft²	Nozzle Sizes: Vapor, in.	Liquid, in.	Steam, in.	Cond., in.	Dimensions: A in.	B in.	C in.	D in.	E ft-in.
16	108	4-11 3/4	132	6	4	4	1 1/2	7 15/16	5 15/16	8	5 3/4	6-1 3/4
20	176	4-11 3/4	215	8	6	4	2	8 15/16	7 15/16	8	5 3/4	6-4 3/4
24	272	4-11 3/4	333	10	6	6	3	9 15/16	7 15/16	9	7 3/4	6-5 3/4
30	431	4-11 3/4	527	12	6	6	3	11 7/16	7 15/16	9	7 3/4	6-7 1/4
36	601	4-11 3/4	735	16	8	8	4	13 3/16	9 15/16	10 3/4	8	6-11
24	272	6-7 3/4	448	10	6	6	3	9 15/16	7 15/16	9	6 3/4	8-1 3/4
30	431	6-7 3/4	710	12	6	6	3	11 7/16	7 15/16	9	6 3/4	8-3 1/4
36	601	6-7 3/4	990	16	8	8	4	13 3/16	9 15/16	10 3/4	8	8-7
42	870	6-7 3/4	1,440	16	10	8	4	17 11/16	10 15/16	10 1/16	6 13/16	9-0 1/2
30	431	9-11 3/4	1,065	12	6	8	3	11 7/16	7 15/16	9	6 3/4	11-7 1/4
36	601	9-11 3/4	1,520	16	8	8	4	13 3/16	9 15/16	10 3/4	6 3/4	11-11
42	870	9-11 3/4	2,180	16	10	8	4	17 11/16	10 15/16	10 1/16	6 13/16	12-4 1/2

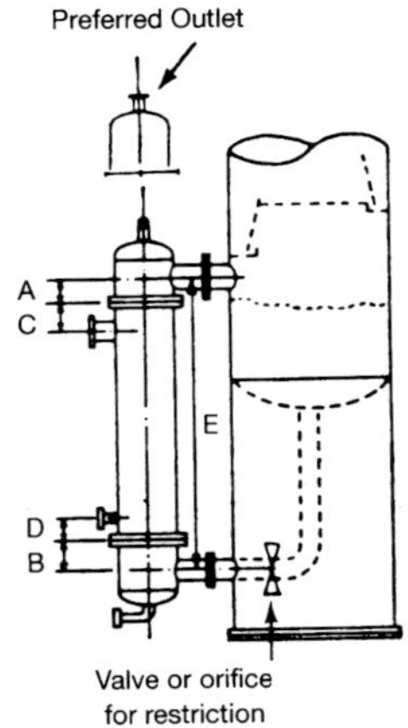

*By permission, D. C. Lee, J. W. Dorsey, G. Z. Moore, and F. D. Mayfield, *Chem. Eng. Prog.*, Vol. 52, No. 4., p. 160 (1956).

***Cross section area of vapor nozzle off to channel must be minimum of 1.25 times total flow area of all tubes.*

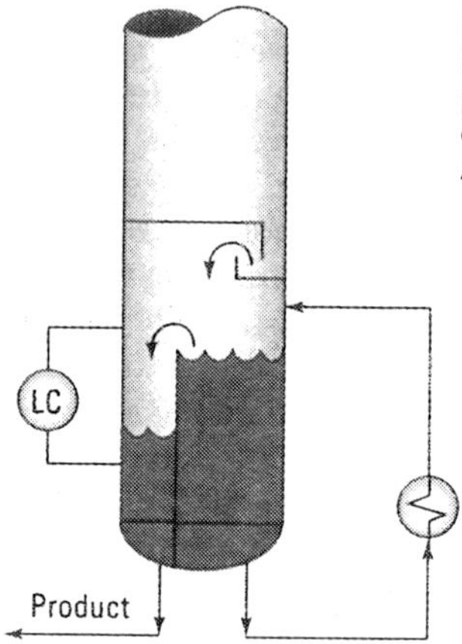

Figure 8-7. Recirculating—Baffled bottom.[26] (Reproduced with permission of the American Institute of Chemical Engineers, © 1997 AIChE, all rights reserved.)

(text continued from page 115)

Reboiler Selection

Figure 8-8 provides an overview of reboiler selection choices. The accompanying notes provide information for a quick or "first-cut" estimate of the appropriate type for a given application. Tables 8-4 and 8-5 provide additional, more detailed, selection data. Table 8-4 gives advantages and disadvantages for all the major reboiler types. Table 8-5 is limited to thermosyphon types.

For reboilers, especially thermosyphon types, "the devil is in the details." The information presented herein is intended for preliminary work. Final design is performed by experienced engineers using detailed design techniques.

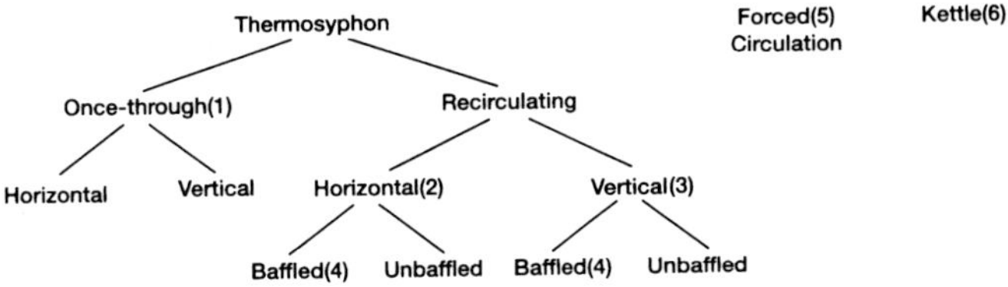

[1] *Preferable to recirculating where acceptable vaporization rates can be maintained (less than 25–30%). This type is chosen when there is a need to minimize exposure of degradable and/or fouling substances to high temperatures.*
[2] *Used for large duties, dirty process, and requires frequent cleaning. The process is usually in the shell side. This type is used in 95% of oil refinery thermosyphon applications.*
[3] *Used for small duties, clean process, and requires only infrequent cleaning. Vaporization is usually less than 30%, but less than 15% if the fractionator pressure is below 50 psig. The viscosity of the reboiler feed should be less than 0.5 cp. Put a butterfly valve in the reboiler inlet piping. This type is used in nearly 100% of chemical plant thermosyphon applications (70% of petrochemical).*
[4] *Greater stability than unbaffled.*
[5] *Usually used where piping pressure drop is high and therefore natural circulation is impractical.*
[6] *Very stable and easy to control. Has no two-phase flow. Allows low tower skirt height. This type is expensive however.*

Figure 8-8. Quick selection guide.

General notes for thermosyphon reboilers:

1. Never use inclined piping for two-phase flow in a process plant. This is particularly true for reboiler return piping. Use only horizontal or vertical runs.
2. If the reboiler heating medium is condensing steam, provide a desuperheater if the superheat is more than 40–60°F.

(text continued on page 125)

Table 8-4
Reboiler Comparison[25]

Type	Advantage	Disadvantage
Kettle	• One theoretical tray • Ease of maintenance • Vapor disengaging • Low skirt height • Handles viscosity greater than 0.5 cp • Ease of control • No limit on vapor load	• Extra piping and space • High cost • Fouls with dirty fluids • High residence time in heat zone for degradation tendency of some fluids • Low residence time surge section of reboiler
Vertical once-through	• One theoretical tray • Simple piping and compact • Not easily fouled • Less cost than kettle	• Difficult maintenance • High skirt height • No control of circulation • Moderate controllability
Vertical natural circulation	• Good controllability • Simple piping and compact • Less cost than kettle	• No theoretical tray • Accumulation of high boiling point components in feed line, i.e., temperature may be slightly higher than tower bottom • Too high liquid level above design could cause reboiler to have less capacity • Fouls easier • Difficult maintenance • High skirt height

Horizontal once-through	• One theoretical tray • Simple piping and compact • Not easily fouled • Lower skirt height than vertical • Less pressure drop than vertical • Longer tubes possible • Ease of maintenance • Less cost than kettle	• No control of circulation • Moderate controllability • High skirt height
Horizontal natural circulation	• Ease of maintenance • Lower skirt height than vertical • Less pressure drop than vertical • Longer tubes possible • Less cost than kettle	• No theoretical tray • Extra space and piping as compared to vertical • Fouls easier as compared to vertical • Accumulation of higher boiling point components in feed line, i.e., temperature may be slightly higher than tower bottom
Forced circulation	• One theoretical tray • Handles high viscous solids-containing liquids • Circulation controlled • Higher transfer coefficient	• Highest cost with additional piping and pumps • Higher operating cost • Requires additional plant area

Reproduced with permission from Hydrocarbon Processing, *Oct. 1992,*

Table 8-5
Thermosyphon Selection Criteria[26]

Factor	Vertical	Vertical	Horizontal	Horizontal
Process side	Tube side	Tube side	Shell side	Shell side
Process flow	Once-through	Circulating	Once-through	Circulating
Heat-transfer coefficient	High	High	Moderately high	Moderately high
Residence time in heated zone	Low	Medium	Low	Medium
ΔT required	High	High	Medium	Medium
Design data	Available	Available	Some available	Some available
Capital cost (total)	Low	Low	Medium	Medium
Plot requirements	Small	Small	Large	Large
Piping needed	Low cost	Low cost	High cost	High cost
Size possible*	Small	Small	Large	Large
Shells	3 maximum	3 maximum	As needed	As needed
Skirt height	High	High	Lower	Lower
Distillation stages	One	Less than one	One	Less than one
Maintenance	Difficult	Difficult	Easy	Easy
Circulation control	None	Possible	None	Possible
Controllability	Moderate	Moderate	Moderate	Moderate
Fouling suitability (process)	Good	Moderate	Good	Moderate
Vaporization range, minimum	5%+	10%+	10%+	15%+
Vaporization range†	25%	25%	25%	25%
Vaporization range, maximum††	35%	35%	35%	35%

**Small = less than 8,000 ft²/shell, large = more than 8,000 ft²/shell.*

†Normal upper limit of standard design range; design for vaporization above this level should be handled with caution.

††Maximum if field data are available.

(text continued from page 121)

PACKED COLUMNS

Packed columns are gaining ground on trayed columns. Lieberman[28] states that based on his design and operating experience, a properly designed packed tower can have 20–40% more capacity than a trayed tower with an equal number of fractionation stages.

Improved packings are being developed, as well as internals and techniques for assuring proper operation. Uniform liquid distribution is imperative for maximum performance. Structured packings are being applied in an expanding range of applications such as high-pressure distillation.

Final design of packed columns should be performed by experts, but the layman is often required to provide preliminary designs for studies. This packed column section provides the information necessary for such estimates.

Packed Column Internals

Preliminary design usually includes a reasonable packing and the use of literature data to determine the height in one or more beds. Standard internals are provided to round out the estimate. See Figure 8-9.

Once packing heights are determined in other sections from HETP (distillation) or K_{GA} (absorption), the height allowances for the internals (from Figure 8-9) can be added to determine the overall column height. Column diameter is determined in sections on capacity and pressure drop for the selected packing (random dumped or structured).

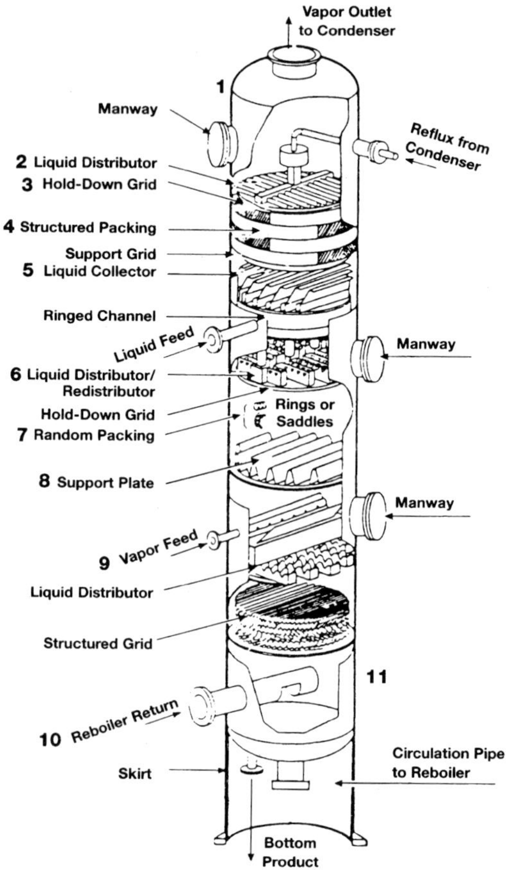

Figure 8-9. Packed column internals.[36] *(Reprinted by special permission from* Chemical Engineering, *March 5, 1984, © 1984 by McGraw-Hill, Inc.)* See table at right.

Note Number	*Section*	*Recommended Height Allowance for Studies*	*Reference*
1.	*Maximum Tower Height*	*175 ft*	*34*
1.	*Top Section*	*4 ft*	*29, 34*
2.	*Liquid Distributor*	*1 ft 8 in.*	*29, 30, 32, 33*
3.	*Hold-Down Grid*	*—*	*30*
4.	*Structured Packing*	*From Text*	*30*
5.	*Liquid Collector*	*3 ft*	*29, 30*
6.	*Liquid Redistributor*	*3 ft wo/manway*	*29, 30*
		4 ft w/manway	*29, 30*
7.	*Random Packing*	*From Text*	*30*
8.	*Support Plate*	*Part of bed*	*30*
9.	*Vapor Feed*	*2 nozzle diameters plus 12 in.*	*29*
10.	*Reboiler Return*	*30 in. High Liquid Level to bed*	*31*
11.	*Bottom Section*	*6 ft*	*29, 34*

Figure 8-9. Continued.

Typical Applications

This subsection will help you select the type of packing to use for your studies and gives typical HETP and HTU values for ballpark estimates when time is short.

Table 8-6 compares packings as an aid to initial selection. For simple applications use Pall Rings for studies as a "tried and true" packing. This will give conservative results when compared with more recent random packings.

Table 8-7 gives typical packing depths for random packing for a variety of applications along with HETP, HTU, and other sizing data.

(text continued on page 131)

Table 8-6
Packing Type Application[1]

Packing	Application Features
Raschig Rings	Earliest type, usually cheaper per unit cost, but sometimes less efficient than others. Available in widest variety of materials to fit service. Very sound structurally. Usually packed by dumping wet or dry, with larger 4–6-inch sizes sometimes hand stacked. Wall thickness varies between manufacturers, also some dimensions; available surface changes with wall thickness. Produce considerable side thrust on tower. Usually has more internal liquid channeling, and directs more liquid to walls of tower. Low efficiency.
Berl Saddles	More efficient than Raschig Rings in most applications, but more costly. Packing nests together and creates "tight" spots in bed which promotes channeling but not as much as Raschig rings. Do not produce as much side thrust, has lower HTU and unit pressure drops with higher flooding point than Raschig rings. Easier to break in bed than Raschig rings.
Intalox Saddles[a] and Other Saddle-Designs	One of most efficient packings, but more costly. Very little tendency or ability to nest and block areas of bed. Gives fairly uniform bed. Higher flooding limits and lower pressure drop than Raschig rings or Berl saddles; lower HTU values for most common systems. Easier to break in bed than Raschig rings, as ceramic.
Pall Rings[b]	Lower pressure drop (less than half) than Raschig rings, also lower HTU (in some systems also lower than Berl saddles), higher flooding limit. Good liquid distribution, high capacity. Considerable side thrust on column wall. Available in metal, plastic and ceramic.
Metal Intalox[a] Hy-Pak[a] Chempak[h]	High efficiency, low pressure drop, reportedly good for distillations.

Spiral Rings	Usually installed as stacked, taking advantage of internal whirl of gas-liquid and offering extra contact surface over Raschig ring, Lessing rings or crosspartition rings. Available in single, double and triple internal spiral designs. Higher pressure drop. Wide variety of performance data not available.
Lessing Rings	Not much performance data available, but in general slightly better than Raschig ring, pressure drop slightly higher. High side wall thrust.
Cross-Partition Rings	Usually used stacked, and as first layers on support grids for smaller packing above. Pressure drop relatively low, channeling reduced for comparative stacked packings. No side wall thrust.
Grid Tile	Available with plain side and bottom or serrated sides and drip-point bottom. Used stacked only. Also used as support layer for dumped packings. Self supporting, no side thrust. Pressure drop lower than most dumped packings and some stacked, lower than some ¼-inch × 1-inch and ¼-inch × 2-inch wood grids, but greater than larger wood grids. Some HTU values compare with those using 1-inch Raschig rings.
Teller Rosette (Tellerette)	Available in plastic, lower pressure drop and HTU values, higher flooding limits than Raschig rings or Berl saddles. Very low unit weight, low side thrust.
Spraypak[c]	Compared more with tray type performance than other packing materials. Usually used in large diameter towers, above about 24-inch dia., but smaller to 10-inch dia. available. Metal only.
Panapak[d]	Available in metal only, compared more with tray type performance than other packing materials. About same HETP as Spraypak for available data. Used in towers 24 inches and larger. Shows some performance advantage over bubble cap trays up to 75 psia in fractionation service, but reduced advantages above this pressure or in vacuum service.

(table continued on next page)

Table 8-6 *Continued*
Packing Type Application[1]

Packing	Application Features
Stedman Packing	Available in metal only, usually used in batch and continuous distillation in small diameter columns not exceeding 24-inches dia. High fractionation ability per unit height, best suited for laboratory work. Conical and triangular types available. Not much industrial data available.
Sulzer, Flexipac, and Similar	High efficiency, generally low pressure drop, well suited for distillation of clean systems, very low HETP.
Goodloe Packing[e] and Wire Mesh Packing	Available in metal and plastic, used in large and small towers for distillation, abosrption, scrubbing, liquid extraction. High efficiency, low HETP, low pressure drop. Limited data available.
Cannon Packing	Available in metal only, low pressure drop, low HETP, flooding limit probably higher than Raschig rings. Not much literature data available. Used mostly in small laboratory or semi-plant studies.
Wood Grids	Very low pressure drop, low efficiency of contact, high HETP or HTU, best used in atmospheric towers of square or rectangular shape. Very low cost.
Dowpac FN-90[f]	Plastic packing of very low pressure drop (just greater than wood slats), transfer coefficients about same as 2-inch Raschig rings. Most useful applications in gas cooling systems or biological trickling filters.
Poly Grid[g]	Plastic packing of very low pressure drop, developed for water-air cooling tower applications.

[a] *Trade name, Norton Co.*
[b] *Introduced by Badische Anilin and Sodafabrik, Ludwigshafen am Rhein*
[c] *Trade name of Denholme Inc., licensed by British Government*
[d] *Trade name of Packed Column Corp.*
[e] *Trade name Packed Column Corp.*
[f] *Trade name of The Dow Chemical Co.*
[g] *Trade name The Fluor Products Co.*
[h] *Trade name Chem-Pro Equip. Corp.*

(text continued from page 127)

Random Dumped Packings

For studies using random dumped packings, determine column diameter and height. Column diameter is determined with use of generalized pressure drop correlations. Column height consists of space occupied by internals (discussed earlier under "Packed Column Internals") and the height of the packing. This subsection gives you methods to determine these items.

First let's discuss column diameter. There is a minimum column diameter for a given sized packing. Table 8-8 shows this relationship.

Table 8-9 shows the maximum liquid loading rates per ft^2 of column diameter. The minimum liquid rate runs 0.5 to 2 gpm/ft^2.

Table 8-10 shows ranges of pressure drop for design. Pressure drop sets the allowable vapor flow rate. The flood pressure drop, for random or structured packings, is given in Reference 40 as:

$$\Delta P_{flood} = 0.115F^{0.7} \tag{8-14}$$

For this equation to apply, the updated packing factors from Reference 30 or 39 must be used.

The pressure drop for a given proposed set of conditions is determined from Figure 8-10 or Figure 8-11, which has a linear Y-axis, so it is easier to interpolate. Table 8-11 gives packing factors (F) to be used with Figures 8-10 and 8-11.

(text continued on page 134)

Table 8-7
Typical Packing Depths[37]

System	L/G lb(/hr-ft²)	Diam. in.	Packing Type	Size, in.
Absorber	8,300/11,000	36	Pall rings	2
L.O.-Top fractionator	3,600/4,700	36	Pall rings	2
L.O.-Bottom fractionator	10,600/5,600	48	Pall rings	2
Deethanizer top	11,000/4,600	18	Pall rings	1½
Deethanizer bottom	19,000/4,500	30	Pall rings	2
Depropanizer top	5,700/4,200	23.4	Pall rings	1½
Depropanizer bottom	5,700/4,200	23.4	Pall rings	1½
Debutanizer top	1,900/3,100	19.5	Pall rings	1½
Debutanizer bottom	1,900/3,100	19.5	Pall rings	1½
Pentane-iso-pentane	2,100/1,900	18	Pall rings	1
Light and heavy naphtha	660/1,250	15	Pall rings	1
	320/600	15	Pall rings	1
	620/1,450	15	Pall rings	1
	260/650	15	Pall rings	1
	370/850	15	Intalox	1
	210/500	15	Intalox	1
	340/800	15	Raschig rings	1
	210/500	15	Raschig rings	1
Iso-octane Toluene	1,970/2,300	15	Pall rings	1
	950/2,100	15	Pall rings	1
	2,100/2,660	15	Pall rings	1
	960/1,200	15	Pall rings	1
	1,110/1,300	15	Pall rings	1
	510/600	15	Pall rings	1
	1,020/1,300	15	Pall rings	1
	470/600	15	Pall rings	1
Gas plant absorber	4,700/6,000	48	Pall rings	2

Table 8-7

Bed depth, ft	HETP, ft	HTU, ft	System press., psia	ΔP, in H_2O/ft pkg	% Overhead
23.0	2.8	-	865	0.55	-
17.0	2.5	-	157	0.12	-
17.0	2.8	-	157	0.30	-
20.0	2.9	-	300	0.20	-
18.0	3.3	-	300	0.30	-
16.0	3.2	-	270	0.30	-
24.0	2.4	-	270	0.30	-
12.0	2.4	-	90	0.12	-
18.0	2.0	-	90	0.12	-
9.0/7.6	1.5	-	Atmos.	0.40	-
10.0	2.00	2.05	100 mm. Hg	1.10	95.0
10.0	3.25	2.50	100 mm. Hg	0.20	95.0
10.0	1.45	1.25	100 mm. Hg	1.75	97.5
10.0	1.45	1.30	100 mm. Hg	0.20	97.5
10.0	2.30	1.90	100 mm. Hg	0.80	93.0
10.0	2.70	2.10	100 mm. Hg	0.22	99.0
10.0	1.95	1.40	100 mm. Hg	1.11	91.6
10.0	2.70	1.97	100 mm. Hg	0.40	96.5
10.0	1.34	1.35	Atmos.	0.70	82.0
10.0	1.90	2.17	Atmos.	0.10	76.0
10.0	0.80	1.02	Atmos.	1.70	84.0
10.0	1.53	1.42	Atmos.	0.15	74.0
10.0	1.34	1.29	100 mm. Hg	1.08	92.5
10.0	1.88	1.81	100 mm. Hg	0.20	87.0
10.0	1.67	1.60	100 mm. Hg	1.14	92.0
10.0	2.07	2.00	100 mm. Hg	0.20	89.0
23.0	2.90	-	900	0.11	92.0 propane absorbed

(text continued from page 131)

Next, to determine packed column height use Table 8-12 for distillation HETP values, leaning towards the high side of the range for studies. For use of K_{GA} values, see Section 9—Absorption. Bed height per packed bed runs up to 20–30 ft for metal or ceramic packings, but plastic packing is usually limited to 24 ft.

Table 8-8
Maximum Packing Size[30]

Nominal Packing Size (in.)	Minimum Column ID (in.)
1	12
1½	18
2	24
3½	42

Table 8-9
Maximum Recommended Liquid Loading for Random Packings[30]

Packing Size (in.)	Liquid Rate (gpm/ft²)
¾	25
1	40
1½	55
2	70
3½	125

Table 8-10
Design Pressure Drop[37]

Service	Pressure Drop (in. H_2O/ ft packed depth)
Absorbers and Regenerators (Non-foaming Systems)	0.25 to 0.40
Absorbers and Regenerators (Moderate Foaming Systems)	0.15 to 0.25
Fume Scrubbers (Water Absorbent)	0.40 to 0.60
Fume Scrubbers (Chemical Absorbent)	0.25 to 0.40
Atmospheric or Pressure Fractionators	0.40 to 0.80
Vacuum Fractionators	0.15 to 0.40

At high liquid viscosities (often occurring in vacuum distillation), the HETP will increase. Table 8-13 shows this relationship.

Structured Packings

Generalizations are not as easy to make for structured packings as for random packings.

See Reference 39 for its large database of structured packing information. One parameter for structured packing

(text continued on page 140)

Table 8-11
Packing Factors (F)—Random Dumped Packings[30]

	Nominal Packing Size (in.)							
	½	⅝	¾	1	1¼	1½	2	3 or 3½
IMTP® Packing (Metal)		51		41		24	18	12
Hy-Pak® Packing (Metal)				45		32	26	16
Super Intalox® Saddles (Ceramic)				60			30	
Super Intalox® Saddles (Plastic)				40			28	18
Intalox Snowflake® (Plastic)							13	
Pall Rings (Plastic)		95		55		40	26	17
Pall Rings (Metal)		81		56		40	27	18
Intalox® Saddles (Ceramic)	200		145	92		52	40	22
Raschig Rings (Ceramic)	580	380	255	179	125	93	65	37
Raschig Rings (1/32-in. Metal)	300	170	155	115				
Raschig Rings (1/16-in. Metal)	410	300	220	144	110	83	57	32
Berl Saddles (Ceramic)	240		170	110		65	45	
Tellerettes (Plastic)				35			24	17

Table 8-12
Separation Efficiency in Standard Distillation Systems[38]

Nominal Packing Size (in.)	HETP (in.)
¾	11 to 16
1	14 to 20
1½	18 to 27
2	22 to 34
3 or 3½	31 to 45

Table 8-13
Effect of Liquid Viscosity on Packing Efficiency[30]

Liquid Viscosity (cps)	Relative HETP (%)
0.22	100
0.35	110
0.75	130
1.5	150
3.0	175

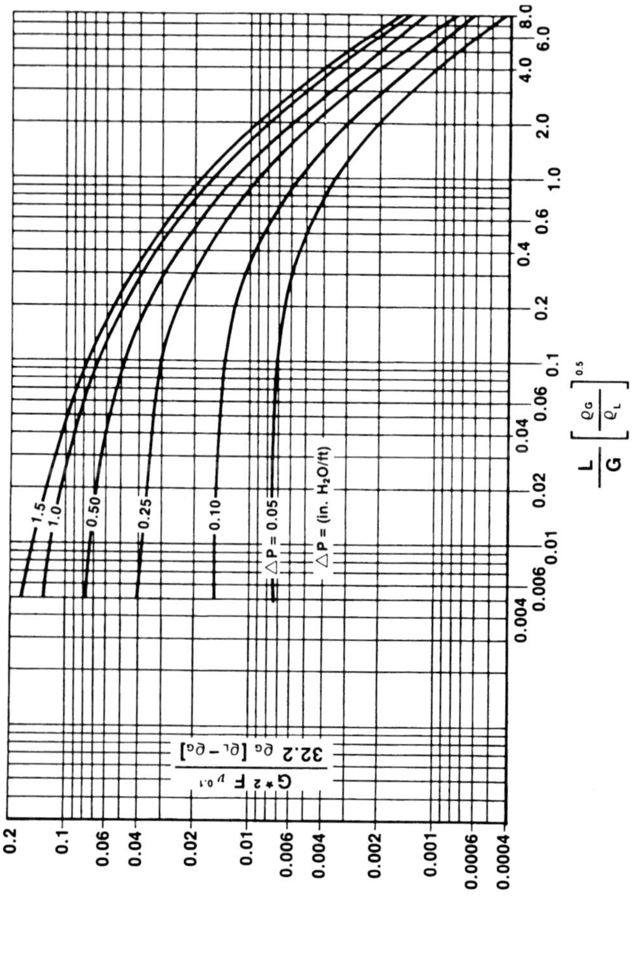

Figure 8-10. Generalized pressure drop correlation.[30]

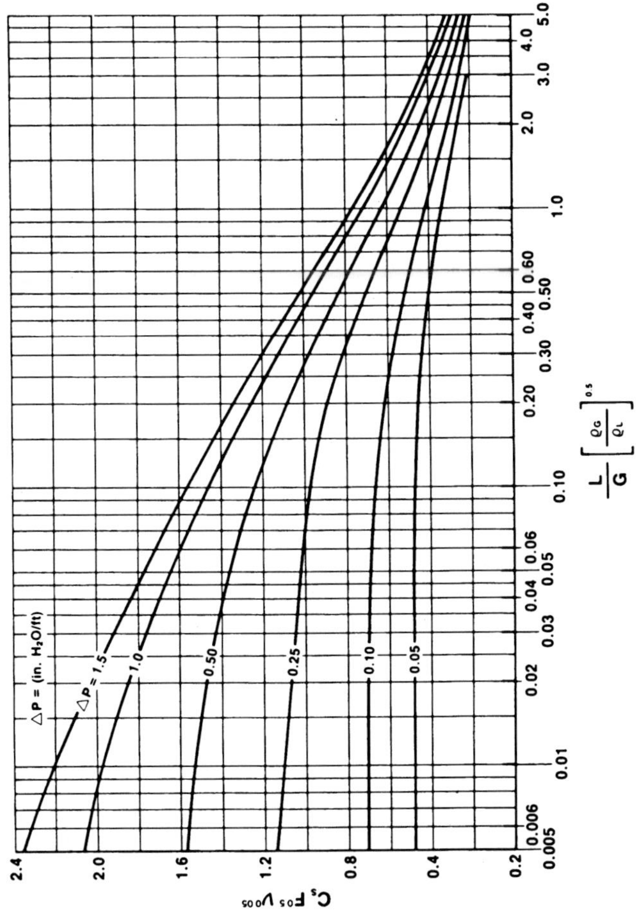

Figure 8-11. Alternate generalized pressure drop correlation.[30]

(text continued from page 135)

recommended in Reference 39 is that the minimum wetting rate is 0.1 to 0.2 gpm/ft^2 compared to 0.5 to 2 gpm/ft^2 for random packings.

The flood pressure drop for structured packing, already shown in the subsection on random dumped packings, is repeated here.[40]

$$\Delta P_{flood} = 0.115F^{0.7}$$

For this equation to apply, the updated packing factors from Reference 30 or 39 must be used.

Nomenclature

C_s or C_p = capacity factor, ft/s
= $U_G\,[\rho_G/(\rho_L - \rho_G)]^{0.5}$
F = packing factor, dimensionless
G = gas mass velocity, lb/ft^2 hr
G* = gas mass velocity, lb/ft^2 sec
HETP = height equivalent to a theoretical stage, in. or ft
HTU = height of a transfer unit, ft
K_{GA} = overall gas mass-transfer coefficient, lb moles/(hr) (ft^2) (atm)
L = liquid mass velocity, lb/ft^2 hr
U_G = gas velocity, ft/sec
ΔP = bed pressure drop, in. H_2O/ft of packing

ρ_G = gas density, lb/ft^3
ρ_L = liquid density, lb/ft^3
ν = kinematic liquid viscosity, centistokes
(centistokes = centipoises/sp.gr.)

REFERENCES

1. Ludwig, E. E., *Applied Process Design for Chemical and Petrochemical Plants,* Vol. 2, 3rd ed. Gulf Publishing Company, Houston, Tex., 1997.
2. Maxwell, J. B., *Data Book on Hydrocarbons,* Van Nostrand, 1965.
3. Norton, *Chemical Process Products,* Norton Company, Chemical Process Products Division.
4. Drickamer, H. G. and J. B. Bradford, "Overall Plate Efficiency of Commercial Hydrocarbon Fractionating Columns," *Trans. AIChE* 39, 319, 1943.
5. O'Connell, H. E., "Plate Efficiency of Fractionating Columns and Absorbers," *Trans. AIChE* 42, 741, 1946.
6. Gunness, *Ind. Eng. Chem.* 29, 1092, 1937.
7. Erbar, J. H. and R. N. Maddox, "Latest Score: Reflux vs. Trays," *Petroleum Refiner,* 40 No. 5, 183, 1961.
8. Fair, J. R. and W. L. Bolles, Monsanto Co., "A Feature Report on Modern Design of Distillation Columns," *Chemical Engineering,* April 22, 1968.
9. Hengstebeck, R. J., *Trans. AIChE* 42, 309, 1946.
10. Underwood, A. J. V., *Trans. Inst. Ch.E.* 10, 112, London, 1932.

11. Souders, M. and G. Brown, "Design of Fractionating Columns—I Entrainment and Capacity," *Ind. Eng. Chem.* 26, 98, 1934.
12. Fenske, M., *Ind. Eng. Chem.* 24, 482, 1932.
13. Smith, B. D., *Design of Equilibrium Stage Processes,* McGraw-Hill, 1963.
14. *GPSA Engineering Data Book,* 10th ed., Gas Processors Suppliers Association.
15. Figure 8, from "Design Data for Thermosiphon Reboilers," by D. C. Lee, J. W. Dorsey, G. Z. Moore, and F. Drew Mayfield, *Chemical Engineering Progress,* Vol. 52, No. 4, pp. 160–164, 1956. Reproduced by permission of the American Institute of Chemical Engineers. © 1956 AIChE.
16. Collins, Gerald K., "Horizontal-Thermosiphon Reboiler Design," *Chemical Engineering,* July 19, 1976.
17. Fair, J. R., "Design Steam Distillation Reboilers," *Hydrocarbon Processing and Petroleum Refiner,* Feb. 1963.
18. Fair, J. R., "Vaporizer and Reboiler Design," parts 1 and 2, *Chemical Engineering,* July 8 and Aug. 5, 1963.
19. Fair, J. R., "What You Need to Design Thermosyphon Reboilers," *Petroleum Refiner,* Feb. 1960.
20. Frank, O. and R. D. Prickett, "Designing Vertical Thermosyphon Reboilers," *Chemical Engineering,* Sept. 3, 1973.
21. Kern, Robert, "How to Design Piping for Reboiler Systems," *Chemical Engineering,* August 4, 1975.

22. Kern, Robert, "Thermosyphon Reboiler Piping Simplified," *Hydrocarbon Processing,* Dec. 1968.
23. Orrell, W. H., "Physical Considerations in Designing Vertical Thermosyphon Reboilers," *Chemical Engineering,* Sept. 17, 1973.
24. Branan, C. R., *The Fractionator Analysis Pocket Handbook,* Gulf Publishing Co., Houston, Tex., 1978.
25. Love, D. L., "No Hassle Reboiler Selection," *Hydrocarbon Processing,*" Oct. 1992.
26. Sloley, A. W., "Properly Design Thermosyphon Reboilers," *Chemical Engineering Progress,* March 1997.
27. McCarthy, A. J. and B. R. Smith, "Reboiler System Design—The Tricks of the Trade," *Chemical Engineering Progress,* May 1995.
28. Lieberman, N. P., *Process Design for Reliable Operations,* 2nd ed., Gulf Publishing Co., Houston, Tex., 1988.
29. Private correspondence, June 2, 1997.
30. Strigle, R. F., *Packed Tower Design and Applications,* 2nd ed., Gulf Publishing Co., Houston, Tex., 1994.
31. Kitterman, Layton, "Tower Internals and Accessories," paper presented at Congresso Brasileiro de Petro Quimiea, Rio de Janeiro, Nov. 8–12, 1976.
32. Frank, Otto, "Shortcuts for Distillation Design," *Chemical Engineering,* March 14, 1977.
33. Graf, Kenneth, "Correlations for Design Evaluation of Packed Vacuum Towers," *Oil and Gas Journal,* May 20, 1985.

34. Dr. Richard Long, Dept. of Chemical Engineering, New Mexico State University, "Guide to Design for Chemical Engineers," handout to students for plant designs, 1994.
35. *GPSA Engineering Data Book,* Gas Processors Suppliers Association, Vol. II, 10th ed.
36. Chen, G. K., "Packed Column Internals," *Chemical Engineering,* March 5, 1984.
37. Eckert, J. S., "Tower Packings—Comparative Performance," *Chemical Engineering Progress,* 59 (5), 76–82, 1963.
38. Branan, C. R., *The Process Engineer's Pocket Handbook,* Vol. 2, Gulf Publishing Co., Houston, Tex., 1983.
39. Kister, H. Z., *Distillation Design,* McGraw-Hill, New York, 1992.
40. Kister, H. Z., "Troubleshoot Distillation and Simulations," *Chemical Engineering Progress,* June 1995.

9
Absorption

A general study of absorption can be confusing because the calculation methods for the two major types are quite different. First, there is hydrocarbon absorption using a lean oil having hydrocarbon components much heavier than the component absorbed from the gas stream. These absorbers may or may not be reboiled. When designing these, use equilibrium vaporization constants (K values) similarly to distillation. Another similarity to distillation is the frequent use of fractionating trays instead of packing. Canned computer distillation programs usually include hydrocarbon absorber options.

The other major type is gas absorption of inorganic components in aqueous solutions. For this type design, use mass transfer coefficients. Packed towers are used so often for this type that its discussion is often included under sections on packed towers. However, in this book it is included in this chapter on absorption.

HYDROCARBON ABSORPTION

Because of its similarity to distillation, many parts of this subject have already been covered, such as

1. Tray efficiency
2. Tower diameter calculations
3. K values

As for distillation, shortcut hand calculation methods exist for hydrocarbon absorption. In distillation, relative volatility (α) values are generated from the K values. For hydrocarbon absorption the K values are used to generate absorption and stripping factors. The 1947 Edmister Method[7], using effective overall absorption and stripping factors and the well known Edmister graphs, is very popular for hand calculations. An excellent write-up on this and the Kremser-Brown-Sherwood Methods are on pages 38–50 of Reference 1.

Edmister Method (1947)

Briefly, the Edmister absorption method (1947) with a known rich gas going to a fixed tower is as follows:

1. Assume theoretical stages and operating temperature and pressure.
2. Knowing required key component recovery E_a, read A_e from Figure 9-1 at known theoretical trays n.

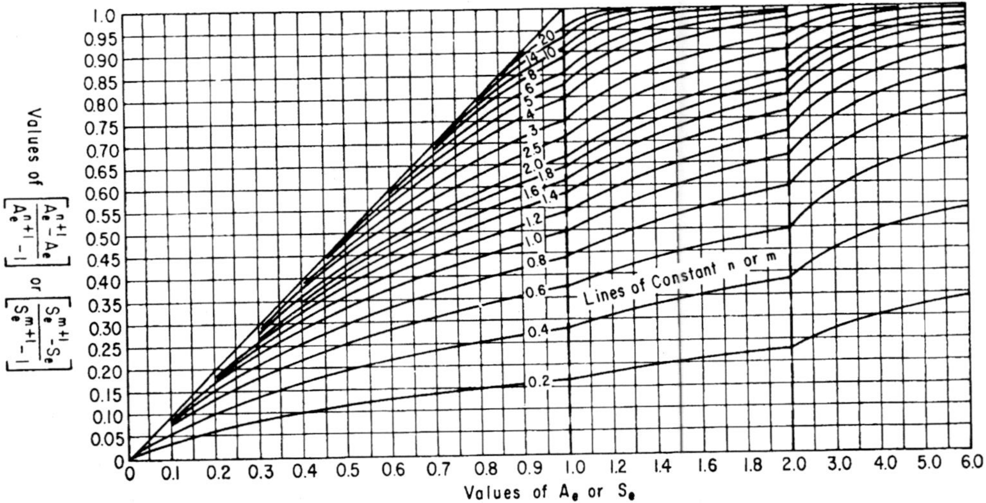

Figure 9-1. This graph shows the absorption and stripping factors, E_a and E_s, versus effective values, A_e and S_e (efficiency functions). (By permission, W. C. Edmister, *Petroleum Engineer*, Sept. 1947 Series to Jan. 1948.)

$$E_a = \frac{A_e^{n+1} - A_e}{A_e^{n+1} - 1} \tag{9-1}$$

$$E_s = \frac{S_e^{m+1} - S_e}{S_e^{m+1} - 1} \tag{9-2}$$

where n = number of theoretical stages in absorber
m = number of theoretical stages in stripper
E_a = fraction absorbed
E_s = fraction stripped
A_e = effective absorption factor
S_e = effective stripping factor

3. Assume
   a. Total mols absorbed
   b. Temperature rise of lean oil (normally 20–40°F)
   c. Lean oil rate, mols/hr

Table 9-1
Tower Balance

Quantity	Symbol	Equation
Rich gas entering at bottom	V_{n+1}	Known
Gas absorbed	ΔV	Assumed
Lean gas leaving absorber	V_1	$V_1 = V_{n+1} - \Delta V$
Gas leaving bottom tray	V_n	$V_n = V_{n+1}(V_1/V_{n+1})^{1/N}$
Gas leaving tray 2 from top	V_2	$V_2 = V_1/(V_1/V_{n+1})^{1/N}$
Lean oil	L_o	Assumed
Liquid leaving top tray	L_1	$L_1 = L_o + V_2 - V_1$
Liquid leaving bottom tray	L_n	$L_n = L_o + \Delta V$

Table 9-2
Tower Temperatures

Temperature	Symbol	Equation
Rich gas inlet	T_{n+1}	Known
Lean oil	T_o	Known
Temperature rise	ΔT	Assumed
Bottom tray	T_n	$T_n = T_{n+1} + \Delta T$
Top tray	T_1	$T_1 = T_n - \Delta T \frac{(V_{n+1} - V_2)}{(V_{n+1} - V_1)}$

4. Use Horton and Franklin's relationship[9] for tower balance in mols/hr. This is shown in Table 9-1.
5. Calculate L_1/V_1 and L_n/V_n.
6. Use the Horton/Franklin Method to estimate tower temperatures. This is shown in Table 9-2.
7. Obtain top and bottom K values.
8. Calculate absorption factors for each component i at the top and bottom

$$A_{Ti} = L_1/K_{1i}\ V_1 \tag{9-3}$$

$$A_{Bi} = L_n/K_{ni}\ V_n \tag{9-4}$$

For stripping factors

$$S_{Ti} = V_1\ K_{1i}/L_1 \tag{9-5}$$

$$S_{Bi} = V_n\ K_{ni}/L_n \tag{9-6}$$

9. Obtain A_{ei} from

$$A_e = [A_B(A_T + 1) + 0.25]^{1/2} - 0.5 \tag{9-7}$$

Similarly

$$S_e = [S_T(S_B + 1) + 0.25]^{1/2} - 0.5 \tag{9-8}$$

10. Read E_{ai} values from Figure 9-1.
11. Calculate mols of each component absorbed.
12. Compare to assumed total mols absorbed and reassume lean oil rate if necessary.

Edmister Method (1957)

Edmister has developed an improved procedure[8] that features equations combining absorption and stripping functions as follows:

$$V_1 = \phi_a V_{n+1} + (1 - \phi_s) L_o \quad \text{(Absorption section)} \tag{9-9}$$

$$L_1 = \phi_s L_{m+1} + (1 - \phi_a) V_o \quad \text{(Stripping section)} \tag{9-10}$$

where L_1 = liquid from bottom stripping tray
L_{m+1} = liquid to top stripping tray
ϕ_a = 1 - E_a, fraction not absorbed
ϕ_s = 1 - E_s, fraction not stripped
V_o = vapor to bottom stripping tray

Other symbols are defined in Tables 9-1 and 9-2. Figure 9-1 and Equations 9-3–9-8 are used as before. V_1 and L_1

are found from Equations 9-9 and 9-10. The improved procedure is better for rigorous solution of complicated absorber designs.

Lean Oil

The selection of lean oil for an absorber is an economic study. A light lean oil sustains relatively high lean oil loss, but has the advantage of high mols/gal compared to a heavier lean oil. The availability of a suitable material has a large influence on the choice. A lean oil 3 carbon numbers heavier than the lightest component absorbed is close to optimum for some applications. In natural gas plant operations, however, the author generally sees a lean oil heavier by about 10–14 carbon numbers.

Presaturators

A presaturator to provide lean oil/gas contact prior to feeding the lean oil into the tower can be a good way to get more out of an older tower. Absorber tray efficiences run notoriously low. A presaturator that achieves equilibrium can provide the equivalent of a theoretical tray. This can easily equal 3–4 actual trays. Some modern canned computer distillation/absorption programs provide a presaturator option.

INORGANIC ABSORPTION

Design of this type absorber quite often involves a system whose major parameters are well defined such as system film control, mass transfer coefficient equations, etc. Reference 1 gives design data for certain well known systems such as

NH_3-Air-H_2O, Cl_2-H_2O, CO_2 in alkaline solutions, etc. Likewise, data for commercially available packings is well documented, such as packing factors, HETP, HTU, etc. Packing parameters are discussed in Chapter 8.

Film Control

The designer must know whether his system is gas or liquid film controlling. For commercial processes this is known. In general an absorption is gas film controlling if essentially all resistance to mass transfer is in the gas film. This happens when the gas is quite soluble in or reactive with the liquid. Reference 1 gives a listing of film control for a number of commercial systems. If a system is essentially all gas or liquid film controlling, it is common practice to calculate only the controlling mass transfer coefficient. Reference 5 states that for gas absorption, the gas mass transfer coefficient is usually used, and for stripping, the liquid mass transfer coefficient is usually used.

Mass Transfer Coefficients

General equations for mass transfer coefficients are given in various references if specific system values are not available. These must, however, be used in conjunction with such things as packing effective interfacial areas and void fractions under operating conditions for the particular packing selected. It is usually easier to find K_{GA} for the packing used with a specific system than effective interfacial area and operation void fraction. Packing manufacturers' data or references, such as Reference 1, can provide specific system K_{GA} or K_{LA} data. Such data are too voluminous for this small handbook.

If K_{GA} values are available for a known system, those of an unknown system can be approximated by

$$K_{GA}(\text{unknown}) = K_{GA}(\text{known})\left[\frac{D_v \text{unknown}}{D_v \text{known}}\right]^{0.56} \tag{9-11}$$

where K_{GA} = gas film overall mass transfer coefficient, lb mols/hr (ft^3) (atm)

D_v = diffusivity of solute in gas, ft^2/hr

Diffusivities

The simplest gas diffusivity relationship is the Gilliland,[1]

$$D_v = 0.0069 \frac{T^{3/2}(1/M_A + 1/M_B)^{1/2}}{P\left(V_A^{1/3} + V_B^{1/3}\right)^2} \tag{9-12}$$

where T = absolute temperature, °R

M_A, M_B = molecular weights of the two gases, A and B

P = total pressure, atm

V_A, V_B = molecular volumes of gases

Height of Overall Transfer Unit

Transfer unit heights are found as follows:

$$H_{OG} = \frac{G_m}{K_{GA} P_{AVE}} \tag{9-13}$$

$$H_{OL} = \frac{L_m}{K_{LA}\rho_L} \tag{9-14}$$

where H_{OG}, H_{OL} = height of transfer unit based on overall gas or liquid film coefficients, ft

G_m, L_m = gas or liquid mass velocity, lb mols/(hr) (ft^2)

K_{GA}, K_{LA} = gas or liquid mass transfer coefficients, consistent units

P_{AVE} = average total pressure in tower, atm

ρ_L = liquid density, lb/ft^3

Number of Transfer Units

For dilute solutions, the number of transfer units N_{OG} is obtained by

$$N_{OG} = \frac{Y_1 - Y_2}{\dfrac{(Y - Y^*)_1 - (Y - Y^*)_2}{\ln \dfrac{(Y - Y^*)_1}{(Y - Y^*)_2}}} \tag{9-15}$$

where $(Y - Y^*)$ = driving force, expressed as mol fractions

Y = mol fraction of one component (solute) at any point in the gas phase

Y^* = mol fraction gas phase composition in equilibrium with a liquid composition, X

X = mol fraction in the liquid at the same corresponding point in the system as *Y*

1, 2 = inlet and outlet of the system, respectively

REFERENCES

1. Ludwig, E E., *Applied Process Design for Chemical and Petrochemical Plants,* Vol. 2, 3rd ed., Gulf Publishing Company, Houston, Tex., 1997.
2. Fair, James R., "Sorption Processes for Gas Separation," *Chemical Engineering,* July 14, 1969.
3. Zenz, F. A., "Designing Gas-Absorption Towers," *Chemical Engineering,* Nov. 13, 1972.
4. *NGPSA Engineering Data Book,* 9th ed., Natural Gas Processors Suppliers Association, 1972.
5. Norton, *Chemical Process Products,* Norton Company, Chemical Process Products Division.
6. Treybal, R. E., *Mass Transfer Operations,* McGraw-Hill Book Co., Inc., NY, 1955.
7. Edmister, W. C., *Petroleum Engineer,* Sept. 1947 series to Jan. 1948.
8. Edmister, W. C., "Absorption and Stripping-Factor Functions for Distillation Calculation by Manual- and Digital-Computer Methods," *AIChE Journal,* June 1957.
9. Horton, G. W. and B. Franklin, "Calculation of Absorber Performance and Design," *Ind. Eng. Chem.* 32, 1384, 1940.

10
Separators And Accumulators

LIQUID RESIDENCE TIME

For vapor/liquid separators, there is often a liquid residence (holdup) time required for process surge. Tables 10-1, 10-2, and 10-3 give various rules of thumb for approximate work. The vessel design method in this chapter under the "Vapor/Liquid Calculation Method" heading blends the required liquid surge with the required vapor space to obtain the total separator volume. Finally, a check is made to see if the provided liquid surge allows time for any entrained water to settle.

Table 10-1
Residence Time for Liquids

Service (times in minutes)	½ Full[1]	L_{HL} to L_{LL} (minimum)[2*]	½ Full[3]	Miscel-laneous
Tower reflux drum	See Table 10-2	5-based on reflux flow	5- to 10-based on total flow	—
Vapor-liquid separators	—	—	3–5	—
Product to storage	Depends on situation	Fractionator O.H. Prod.—2	—	—
Product to heat exchanger along with other streams	—	Fractionator O.H. Prod. —5	—	—
Product to heater	—	Fractionator O.H. Prod. —10	—	—
Furnace surge drums	—	—	—	10 min. 20 max.
Tower bottoms FRC control	—	—	—	5 min. 10 max.
LC control	—	—	—	3 min.

* *This article deals only with reflux drums. Use only the larger vessel volume determined. Do not add two volumes such as reflux plus product. If a second liquid phase is to be settled, additional time is needed. For water in hydrocarbons, an additional 5 minutes is recommended.*

Table 10-2
Liquid Residence Rules of Thumb for Reflux Drums

½ Full Minutes[1]

Factor	Reflux Control	Instrument Factor		Labor Factor*		
Base reflux drum rules of thumb		w/ Alarm	w/o Alarm	G	F	P
	FRC	½	1	2	3	4
	LRC	1	1½	2	3	4
	TRC	1½	2	2	3	4

The above are felt to be conservative. For tight design cut labor factors by 50%. Above based on gross overhead.

	Situation	Factor
Multipliers for overhead product portion of gross overhead depending on operation of external equipment receiving the overhead product	Under good control	2.0
	Under fair control	3.0
	Under poor control	4.0
	Feed to or from storage	1.25

	Situation	Factor
Multipliers for gross overhead if no board mounted level recorder	Level indicator on board	1.5
	Gauge glass at equipment only	2.0

**Labor factors are added to the instrument factors G = good, F = fair, P = poor.*

Table 10-3
Liquid Residence Rules of Thumb for Proper Automatic Control of Interface Level[3, 4]

Flow (gpm)	Reservoir Capacity (gal/in. of depth)
100	2
200	3
600	4
800	8
1,000	10
1,500	15
2,000	24

VAPOR RESIDENCE TIME

For vapor/liquid separators, this is usually expressed in terms of maximum velocity that is related to the difference in liquid and vapor densities. The standard equation is

$$U_{vapor\ max} = K[(\rho_L - \rho_v)/\rho_v]^{0.5}$$

where U = velocity, ft/sec
ρ = density of liquid or vapor, lb/ft^3
K = system constant

Figure 10-1 relates the K factor for a vertical vessel (K_v) to:

$$W_L/W_v\ (\rho_v/\rho_L)^{0.5}$$

where W = liquid or vapor flow rate, lb/sec

For a horizontal vessel $K_H = 1.25K_v$.

Figure 10-1 is based upon 5% of the liquid entrained in the vapor. This is adequate for normal design. A mist eliminator can get entrainment down to 1%.

An equation has been developed for Figure 10-1 as follows:

$X = \ln[(W_L/W_v)(\rho_v/\rho_L)^{0.5}]$

$Y = K_v$ (Remember $K_H = 1.25K_v$)

$Y = \mathrm{EXP}\,(A + BX + CX^2 + DX^3 + EX^4 + FX^5)$

$A = -1.942936$

$B = -0.814894$

$C = -0.179390$

$D = -0.0123790$

$E = 0.000386235$

$F = 0.000259550$

VAPOR/LIQUID CALCULATION METHOD

A vessel handling large amounts of liquid or a large liquid surge volume will usually be horizontal. Also, where water must be separated from hydrocarbon liquid, the vessel will be horizontal. A vessel with small surge volume such as a compressor knockout drum will usually be vertical.

Vertical Drum

This method uses the separation factor given in the section titled "Vapor Residence Time." The first three steps

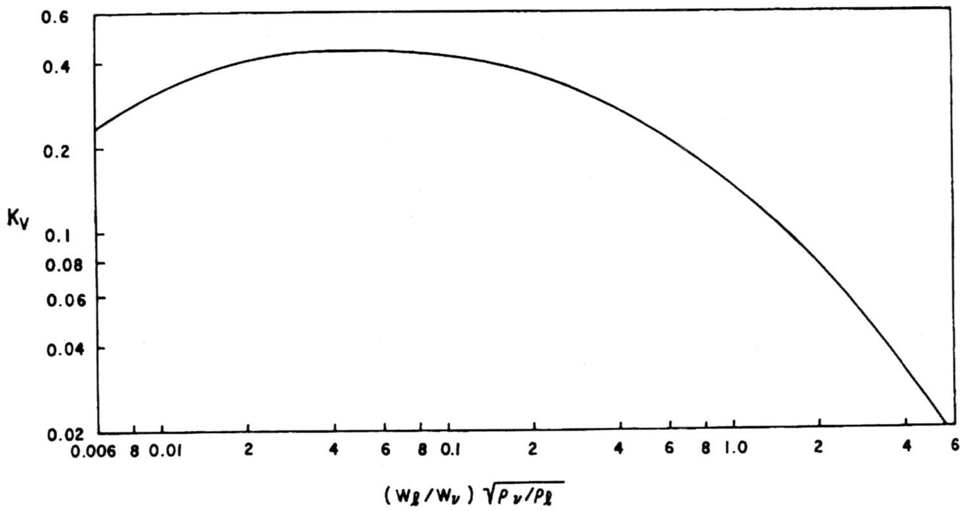

Figure 10-1. Design vapor velocity factor for vertical vapor-liquid separators at 85% of flooding.

use equations and a graph (or alternate equation) in that section to get K_v and $U_{vapor\ max}$. Nomenclature is explained there.

1. Calculate separation factor = $W_L/W_v\ (\rho_v/\rho_L)^{0.5}$
2. Get K_v from graph or equation
3. Calculate $U_{vapor\ max}$
4. Calculate minimum vessel cross section
5. Set diameter based on 6-in. increments
6. Approximate the vapor-liquid inlet nozzle based on the following criteria:

   (U_{max}) nozzle $= 100/\rho_{mix}{}^{0.5}$
   (U_{min}) nozzle $= 60/\rho_{mix}{}^{0.5}$

   where U = max, min velocity, ft/sec
   ρ_{mix} = mixture density, lb/ft^3

7. Sketch the vessel. For height above center line of feed nozzle to top seam, use 36 in. + ½ feed nozzle OD or 48 in. minimum. For distance below center line of feed nozzle to maximum liquid level, use 12 in. + ½ feed nozzle OD or 18-in. minimum.
8. Select appropriate full surge volume in seconds. Calculate the required vessel surge volume.

   $V = Q_L$ (design time to fill), ft^3

where Q_L = liquid flow, ft^3/sec

9. Liquid height is

$$H_L = V\,(4/\Pi D^2),\ ft$$

10. Check geometry. Keep

$$(H_L + H_v)/D$$

between 3 and 5, where H_v is vapor height in feet.

For small volumes of liquid, it may be necessary to provide more liquid surge than is necessary to satisfy the $L/D > 3$. Otherwise this criteria should be observed. If the required liquid surge volume is greater than that possible in a vessel having $L/D < 5$, a horizontal drum must be provided.

Horizontal Drum

This method is a companion to the vertical drum method.

1. Calculate separation factor.
2. Look up K_H.
3. Calculate $U_{vapor\ max}$.
4. Calculate required vapor flow area.

$$(A_v)_{min} = Q_v/U_{vapor\ max},\ ft^2$$

5. Select appropriate design surge time and calculate full liquid volume. The remainder of the sizing procedure is done by trial and error as in the following steps.
6. When vessel is at full liquid volume,

$$(A_{total})_{min} = (A_v)_{min} / 0.2$$

$$D_{min} = \sqrt{4(A_{total})_{min} / \Pi}, \text{ ft}$$

7. Calculate vessel length.

$$L = \frac{\text{full liquid volume}}{(\Pi / 4)D^2}$$

$D = D_{min}$ to the next largest 6 inches

8. If $5 < L/D < 3$, resize.

If there is water to be settled and withdrawn from hydrocarbon, the water's settling time requirement needs to be checked. The water settling requirement, rather than other process considerations, might set the liquid surge capacity. Therefore, the liquid surge capacity we have previously estimated from tables might have to be increased.

Here is a quick check for water settling.

1. Estimate the water terminal settling velocity using:

$$U_T = 44.7 \times 10^{-8} (\rho_w - \rho_o) F_s/\mu_o$$

where U_T = terminal settling velocity, ft/sec
F_s = correction factor for hindered settling

ρ_w, ρ_o = density of water or oil, lb/ft^3
μ_o = absolute viscosity of oil, lb/ft-sec

This assumes a droplet diameter of 0.0005 ft. F_s is determined from:

$F_s = X^2/10^{1.82\,(1-x)}$

where X = vol. fraction of oil

2. Calculate the modified Reynolds number, Re from:

$Re = 5 \times 10^{-4}\ \rho_o U_T/\mu_o$ (usually < 1.0)

This assumes a droplet diameter of 0.0005 ft

3. Calculate U_S/U_T from:

$U_S/U_T = A + B \ln Re + C \ln Re^2 + D \ln Re^3 + E \ln Re^4$

where U_S = Actual settling velocity, ft/sec
A = 0.919832
B = –0.091353
C = –0.017157
D = 0.0029258
E = –0.00011591

4. Calculate the length of the settling section (Figure 10-2), as:

$$L = hQ/AU_S$$

where L = length of settling zone, ft
h = height of oil, ft
Q = flow rate, ft^3/sec
A = cross-sectional area of the oil settling zone, ft^2

This allows the water to fall out and be drawn off at the bootleg before leaving the settling section.

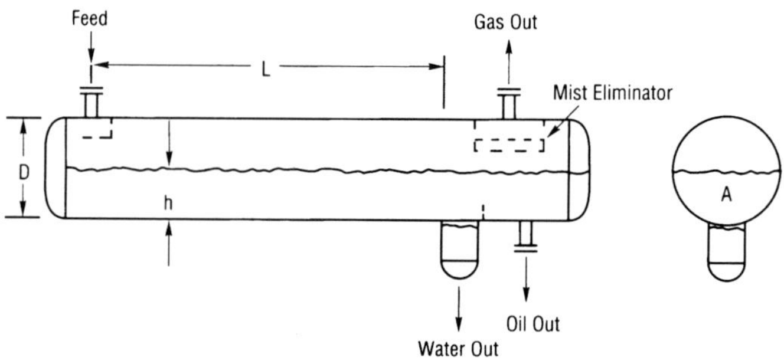

Figure 10-2. Typical vapor/liquid separator.

LIQUID/LIQUID CALCULATION METHOD

R. L. Barton[6] provides the following quick method for sizing liquid-liquid phase separators empirically.

The separation of mixtures of immiscible liquids is an important chemical engineering operation. This empirical design has proven satisfactory for many phase separations.

1. Calculate holdup time with the formula

$$T = 0.1\ [\mu/(\rho_b - \rho_t)]$$

where T = holdup time, hours
μ = viscosity of the continuous phase, cp
ρ_b = sp. gr. of the bottom phase
ρ_t = sp. gr. of the top phase

2. Assign a length-to-diameter ratio of 5, and size a tank to accommodate the required holdup time.
3. Provide inlet and outlet nozzles at one end, and an internal flat cone (see Figure 10-3).

While this design procedure is empirical, there is some rationale behind it. The relation between viscosity and specific gravity difference of the phases corresponds to those of the equations for terminal settling velocity in the Stokes law region and free-settling velocity of isometric particles. Also, the dimensions of the tank and cone recognize that the shape of turbulence created by nozzles discharging into liquids spreads at an angle whose slope is about 1 to 5.

This design is not good for emulsions.

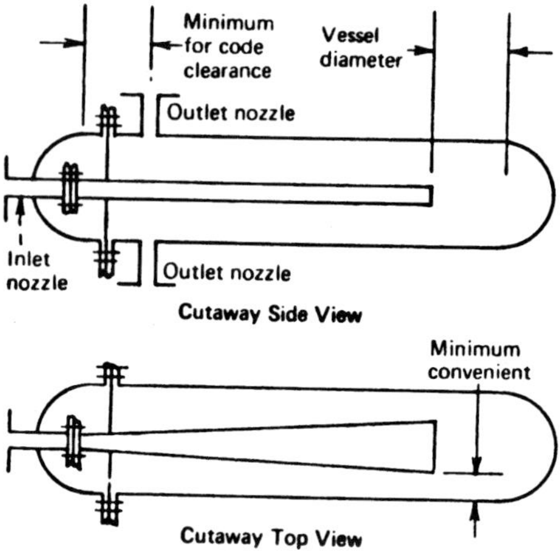

Notes:
1. For large vessels, a manway and internally dismanted cone may be used where more economical.
2. Inlet and outlet nozzles should be sized for pump discharge.
3. Gauge glass and level instruments should be located at inlet-outlet end.
4. Mechanical design should suit economy under operating conditions.

Figure 10-3. Recommended design for a liquid-liquid separator.[6]

PRESSURE DROP

Pressure Drop Across Mist Eliminator

Use 1-in. H_2O pressure drop.

Pressure Drop Entering Plus Leaving Vessel

One velocity head for inlet and one half for outlet pipe velocity is close.

VESSEL THICKNESS

Equation 10-1 gives the required vessel thickness based on inside vessel radius. Equation 10-2 gives the thickness based on outside radius.

$$T = \frac{Pr_i}{SE - 0.6P} + C \qquad (10\text{-}1)$$

$$T = \frac{Pr_o}{SE + 0.4P} + C \qquad (10\text{-}2)$$

where P = pressure, psig
r = radius, in.
S = allowable stress, psi
E = weld efficiency, fraction. Use 0.85 for initial work.
C = corrosion allowance, in.

REFLUX DRUMS

Sigales[2] has provided an optimizing method for reflux drums. Here are some useful comments from that article.

1. Reflux drums are usually horizontal because the liquid load is important.

2. When a small quantity of a second liquid phase is present, a drawoff pot (commonly called a bootleg) is provided to make separation of the heavy liquid (frequently water) easier. The pot diameter is ordinarily determined for heavy phase velocities of 0.5 ft/min. Minimum length is 3 ft for level controller connections. Minimum pot diameter for a 4–8-ft diameter reflux drum is 16 in. For reflux drums with diameters greater than 8 ft, pot diameters of at least 24 in. are used. The pot must also be placed at a minimum distance from the tangent line that joins the head with the body of the vessel.
3. The minimum vapor clearance height above high liquid level is 20% of drum diameter. If possible this should be greater than 10 inches.

VESSEL DESIGN TIPS

The process engineer gets involved in many mechanical aspects of vessel design such as thickness, corrosion allowance, and internals. Here are some pitfalls to watch for along the way:

1. Be sure to leave sufficient disengaging height above demisters[7]; otherwise, a healthy derate must be applied.
2. For liquid/liquid separators, avoid severe piping geometry that can produce turbulence and homogenization. Provide an inlet diffuser cone and avoid shear-producing items, such as slots or holes.

3. Avoid vapor entry close to a liquid level. Reboiler vapor should enter the bottom of a fractionator a distance of at least tray spacing above high liquid level. Tray damage can result if the liquid is disturbed.
4. Avoid extended nozzles or internal piping that the operator cannot see, if at all possible.
5. Make sure items such as gauge glasses, level controls, or pressure taps do not receive an impact head from an incoming stream.
6. Use a close-coupled ell for drawoff from gravity separators to eliminate backup of hydraulic head.
7. Check gravity decanters for liquid seal and vapor equalizing line (syphon breaker).
8. For gauge glasses, it is good to have a vent at the top as well as a drain at the bottom. These should be inline for straight-through cleaning.

GAS SCRUBBERS

Fair Separation

$$G = 900\sqrt{\rho_V(\rho_L - \rho_V)}$$

Good Separation

$$G = 750\sqrt{\rho_V(\rho_L - \rho_V)}$$

where G = allowable mass velocity, lb/hr ft^2
ρ = density, lb/ft^3

F-FACTOR GUIDELINES FOR DISENGAGING

The F-factor is defined as $F = u\,(\rho)^{0.5}$,

where F = factor for allowable velocity
u = velocity, ft/sec
ρ = density, lb/ft^3

For quick estimates the following F-factor guidelines are proposed:

Situation	F-Factor
Fractionating column total cross-section vapor velocity	1.0–1.5
Sieve tray hole velocity to avoid weeping	>12
Disengaging equipment for liquid/vapor separation	<6

ESTIMATING THE SEPARATION

Given:

Comp	X	Temp	Other Data
C_2	.10	70°F	A vapor pressure chart
C_3	.30		
nC_4	.60		
	1.00		

The use of shortcut methods to estimate vapor/liquid equilibria is best illustrated by the following example.

Find:

- Mixture bubble point pressure
- Component K = Y/X values
- For flash at 70°F and 80 psia:
    - Vapor and liquid produced
    - Vapor and liquid compositions
    - Vapor and liquid mol wts
    - Vapor and liquid densities
    - Heat of vaporization

Comp	X	VP, psia	pp, psia	Y	K =Y/X
C_2	.10	600	60.0	.4926	4.9
C_3	.30	140	42.0	.3448	1.15
nC_4	.60	33	19.8	.1626	0.27
	1.00		121.8	1.0000	

Mixture bubble point pressure and K = Y/X values:

Use Raoult's Law: pp = X(VP)

Bubble point pressure at 70°F = 122 psia

Flash at 70°F and 80 psia:

K = Y/X values can also be found another way by Raoult's Law:

Because pp = X (VP) and

pp = TP (Y) then
VP(X) = TP (Y) so
VP/TP = Y/X = K

This route to K values is handy for flash calculations as follows:

Comp	N	VP psia	K = VP/TP	L = 0.6 $X_i = N_i/(L+VK_i)$
C_2	.10	600	7.5	.0278
C_3	.30	140	1.75	.2308
nC_4	.60	33	0.4125	.7843
	1.00			1.0429

L = 0.7 X_i	L = 0.8 X_i	L = 0.73 X_i	L= 0.72 X_i	L = 0.725 X_i
.0339	.0435	.0363	.0355	.0359
.2449	.2609	.2495	.2479	.2487
.7284	.6799	.7131	.7181	.7156
1.0072	.9843	.9989	1.0015	1.0002

At L = 0.725

			Normalized Compositions	
Comp	X	Y = KX	X	Y
C_2	.0359	.2693	.0359	.2694
C_3	.2487	.4352	.2487	.4353
nC_4	.7156	.2952	.7154	.2953
	1.0002	.9997	1.0000	1.0000

Liquid mol wt

Comp	MW	Mols.	lb
C_2	30	.0359	1.08
C_3	44	.2487	10.94
nC_4	58	.7154	41.49
		1.0000	53.50

Vapor mol wt

Comp	MW	Mols.	lb
C_2	30	.2694	8.1
C_3	44	.4353	19.2
nC_4	58	.2953	17.1
		1.0000	44.4

We will use the method in the *GPSA Data Book* that assumes L and calculates $X_i = N_i/(L + VK_i)$. Iterations continue until $X_i = 1.0$.

Vapor Density (70°F, 80 psia)

$$44.4 \text{ lb}/76 \text{ ft}^3 \times (460 + 76)/(460 + 70) \times 80/76 = 0.62 \text{ lb/ft}^3$$

See the subsection "Approximate Physical Properties" for the origin of 76 ft^3/lb mol at 76°F and 76 psia.)

Liquid Density

Because the liquid produced will be at its bubble point use 4 lb/gal as shown in the subsection "Approximate Physical Properties."

Heat of Vaporization

Use the vapor composition and the Clausius/Clapeyron equation in reverse to roughly estimate ΔH_{vap}. This requires two points on the vapor pressure chart.

$$\ln (P/P_o) = (\Delta H_{vap}/R)\,(1/T_o - 1/T)$$

Use vapor pressure points for a fictitious component having the mixture molecular weight. This is better than calculating ΔH_{vap} for individual components. Invariably, a light component or two will be near their critical points and give wild results, off by more than 100%.

Because the vapor mol wt is 44.7, we will use C_3 as our "fictitious component."

Comp	P	P_o	T	T_o	R
C_3	200 psia	80 psia	100°F 560°R	40°F 500°R	1.987 Btu/lb mol°R use 2

$\ln P/P_0 = .9163$; $1/T_o - 1/T = .000214$

$\Delta H_{vap} = 8{,}564$ Btu/lb mol = 191 Btu/lb

The *GPSA Data Book*[10] Figures 24-11 and 24-16 give:

$$\begin{aligned} H_{vapor} &= 152 \text{ Btu/lb} \\ H_{liquid} &= \underline{-(-19)} \\ \Delta H_{vap} & 171 \end{aligned}$$

Clausius Clapeyron gave results 12% higher.

Nomenclature

Mol Fractions

N = in flash feed
X = in liquid
Y = in vapor

Pressures, psia

P = pressure at a selected point on vapor pressure chart
P_0 = pressure at another selected point on vapor pressure chart
pp = partial pressure of a component in vapor
TP = total pressure of the system
VP = vapor pressure of a component

Temperatures, °R

T = temperature of a selected point on vapor pressure chart
T_0 = temperature at another selected point on vapor pressure chart

Quantities, lb mols/hr

L = liquid
V = vapor

Other

i = subscript indicating individual component
K = Y/X
R = gas constant, Btu/lb mol °R
ΔH_{vap} = latent heat of vaporization, Btu/lb mol or Btu/lb

REFERENCES

1. Watkins, R. N., "Sizing Separators and Accumulators," *Hydrocarbon Processing,* Nov. 1967.
2. Sigales, B., "How to Design Reflux Drums," *Chemical Engineering,* March 3, 1975.
3. Younger, A. H., "How to Size Future Process Vessels," *Chemical Engineering,* May 1955.
4. Anderson, G. D., "Guidelines for Selection of Liquid Level Control Equipment," Fisher Controls Company.
5. Branan, C. R., *The Process Engineer's Pocket Handbook,* Vol. 1, Gulf Publishing Co., Houston, Tex., 1976.
6. Barton, R. L., "Sizing Liquid-Liquid Phase Separators Empirically," *Chemical Engineering,* July 8, 1974.
7. "Performance of Wire Mesh Demisters®," Bulletin 635, Otto H. York Co., Inc.
8. Branan, C. R., *The Process Engineer's Pocket Handbook,* Vol. 2, Gulf Publishing Co., Houston, Tex., 1983.
9. Equations were generated using FLEXCURV, Gulf Publishing Co.
10. *GPSA Engineering Data Book,* Gas Processors Suppliers Association, Vol. 2, 10th ed.

11 Refrigeration

TYPES OF SYSTEMS

Table 11-1 shows the three most used refrigeration systems and approximate temperature ranges.

ESTIMATING HORSEPOWER PER TON

Figure 11-1 is a quick, but accurate, graph[2] that shows the design engineer how much horsepower is required for mechanical refrigeration systems, using the most practical refrigerant for the desired temperature range.

Example. A water-cooled unit with an evaporator temperature of –40°F will require 3 hp/ton of refrigeration. A ton of refrigeration is equal to 12,000 Btu/hr.

Here are equations for these curves in the form:

$$y = A + Bx + Cx^2 + Dx^3 + Ex^4$$

Table 11-1
Types of Refrigeration Systems

	Approx. Temp. Range, °F	Refrigerant
1. Steam—Jet	35° to 70°	Water
2. Absorption		
Water—Lithium Bromide	40° to 70°	Lithium Bromide solution
Ammonia	–40° to +30°	Ammonia
3. Mechanical Compression (Reciprocating or centrifugal)	–200° to +40°	Ammonia, halogenated hydrocarbons, propane, ethylene, and others

The most common light hydrocarbon refrigerant cooling temperature ranges are:

Methane	–200 to –300°F
Ethane and ethylene	–75 to –175°F
Propane and propylene	+40 to –50°F

where y = hp/ton refrigeration
x = evaporator temperature, °F

Condenser temperature

°F	A	B	C	D	E
105	1.751	–2.686e-2	1.152e-4	3.460e-8	1.320e-9
120	2.218	–2.882e-2	1.036e-4	3.029e-7	3.961e-9

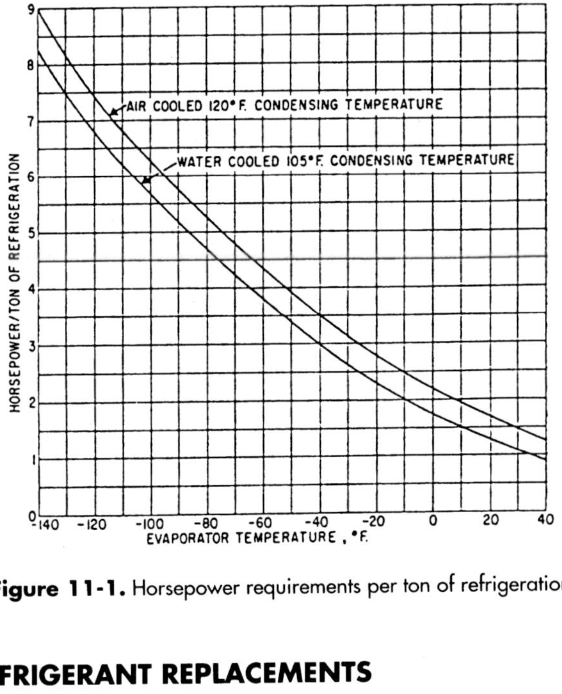

Figure 11-1. Horsepower requirements per ton of refrigeration.

REFRIGERANT REPLACEMENTS

Most countries are phasing out certain refrigerants to lessen damage to the ozone layer. The chemicals being phased out are chloroflurocarbons (CFCs) and hydrochloroflurocarbons (HCFCs). Replacements are hydroflurocarbons (HFCs) and certain blends. The DuPont web site (www.dupont.com) gives the handy Table 11-2 of recommended replacement refrigerants for various applications.

DuPont also provides a computer program to simulate performance of different refrigerants in your system. It is called the "DuPont Refrigeration Expert" (DUPREX).

Table 11-2
DuPont Refrigerants

Application	CFC/HCFC Refrigerants	Retrofit	New System
Air conditioning in buildings and industrial temperature control	R-11	SUVA® 123	SUVA® 123
	R-12	SUVA® 134a SUVA® MP39	SUVA® 134a
	R-22	SUVA® 9000	SUVA® 9000 SUVA® 9100
Split and window A/C systems	R-22	SUVA® 9000	SUVA® 9000 SUVA® 9100
Air and marine A/C systems	R-114, R-12B1	SUVA®124	SUVA®124
Automobile air conditioning	R-12	SUVA®134a SUVA® MP52	SUVA®134a
Fresh food storage, above 0°C	R-12	SUVA® MP39	SUVA® 134a
Domestic refrigerators, drink coolers, commercial and restaurant non-frozen chilled foods storage	R-12	SUVA® MP39	SUVA® 134a

Application	CFC/HCFC Refrigerants	Retrofit	New System
Light commercial refrigeration	R-22	SUVA® HP62	SUVA® HP62
Frozen Food storage below −18°C	R-502	SUVA® HP80 SUVA® HP81	SUVA® HP62
Refrigerated Transport	R-12	SUVA® MP66	SUVA® 134a
Low Temperature Transport	R-502	SUVA® HP80	SUVA® HP62
Medium Temperature commercial refrigeration	R-12, R-500	SUVA® MP39 SUVA® MP66	SUVA® 134a
Low to Medium Temperature commercial refrigeration	R-22	SUVA® HP62	SUVA® HP62
Very low temperature	R-13B1	SUVA® 9100	SUVA®9100
	R-13	HFC-23	HFC-23
	R-503	SUVA®95	SUVA®95

**The choice of a SUVA® refrigerant will depend on the application as well as on the type of CFC to replace.*
This table is intended as a guide, to cover the situations most likely to be encountered. The equipment owner may request the use of the permanent or "New System" refrigerant in a retrofit, usually to ensure the lowest ODP. Retrofitting directly to a SUVA® HFC Refrigerant can of course be done, but takes longer and costs more than the simple change to a SUVA® Blend.

ETHYLENE/PROPYLENE CASCADED SYSTEM

The following information was used in olefin plant case studies to determine whether the ethylene/propylene cascaded refrigeration systems had enough horsepower for various plant operations. The propylene was condensed against cooling water at 110°F and the ethylene was condensed against propylene at –20°F. For comparison, the horsepower requirements for each refrigerant alone are also shown.

The data is presented in the following equation form:

$$y = A + Bx + Cx^2$$

where y = hp/ton refrigeration
x = evaporator temperature, °F

Example. With the example cascaded system at an evaporator temperature of –100°F, the horsepower requirement is 6.2 hp/ton refrigeration. A ton of refrigeration is equal to 12,000 Btu/hr.

Propylene alone at 110°F condensing temperature

A	B	C	Range, °F
1.98	−1.991e-2	1.691e-5	–40 to 110

Ethylene alone at −20°F condensing temperature

A	B	C	Range, °F
4.67	−1.256e-2	7.778e-5	−150 to −60

Cascade with ethylene condensing at −20°F and propylene at 110°F

A	B	C	Range, °F
3.47	−2.289e-2	4.444e-5	−150 to −60

STEAM JET TYPE UTILITIES REQUIREMENTS

Steam and cooling water requirements for barometric steam jet refrigeration units are shown in Figures 11-2–11-5 for given available cooling water temperature and delivered chilled water temperature. The graphs are for 100 psig motivating steam. For 30–50 psig steam, the quantity required will increase by a factor of about 2 for 40°F chilled water and a factor of 1.5 for 55°F chilled water.

Example. To produce 20 tons of refrigeration while delivering 50°F chilled water, the steam consumption depends upon the quantity and temperature of the cooling water. If there are 140 gpm of 85°F cooling water the y-axis is 7 gpm/ton of refrigeration. The steam consumption on the x-axis is about 17 lb/hr steam per ton of refrigeration. A ton of refrigeration is equal to 12,000 Btu/hr.

(text continued on page 191)

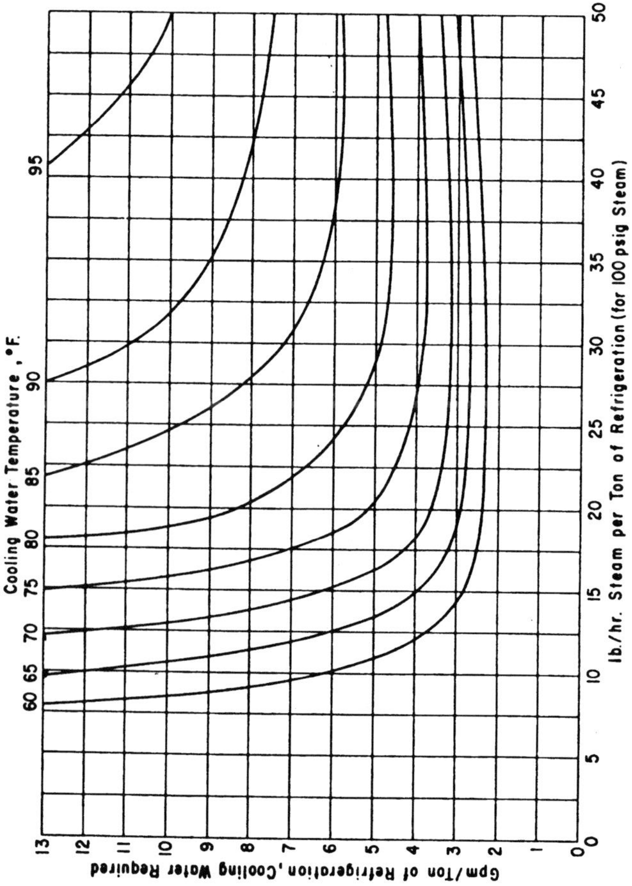

Figure 11-2. lb/hr steam per ton of refrigeration (for 100 psig steam) for 40°F chilled water.

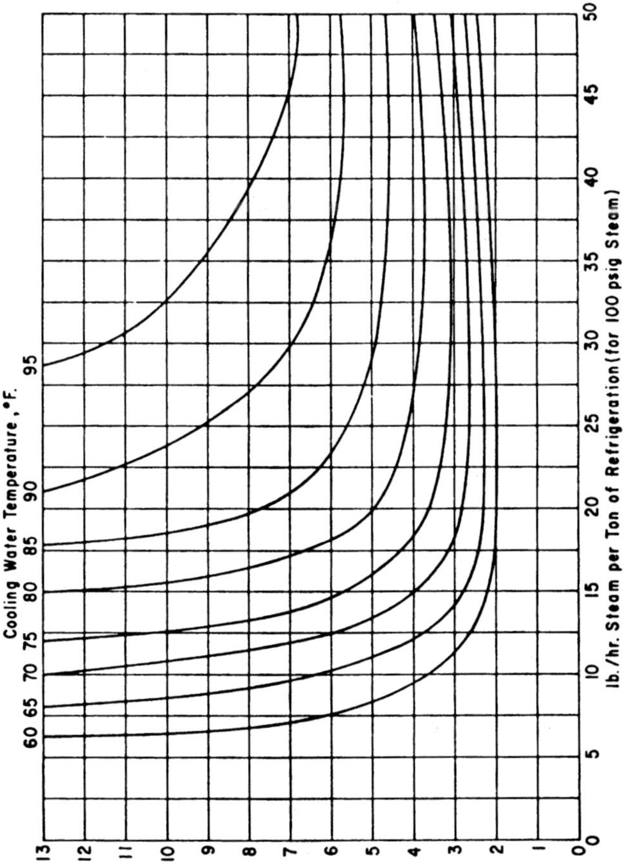

Figure 11-3. lb/hr steam per ton of refrigeration (for 100 psig steam) for 45°F chilled water.

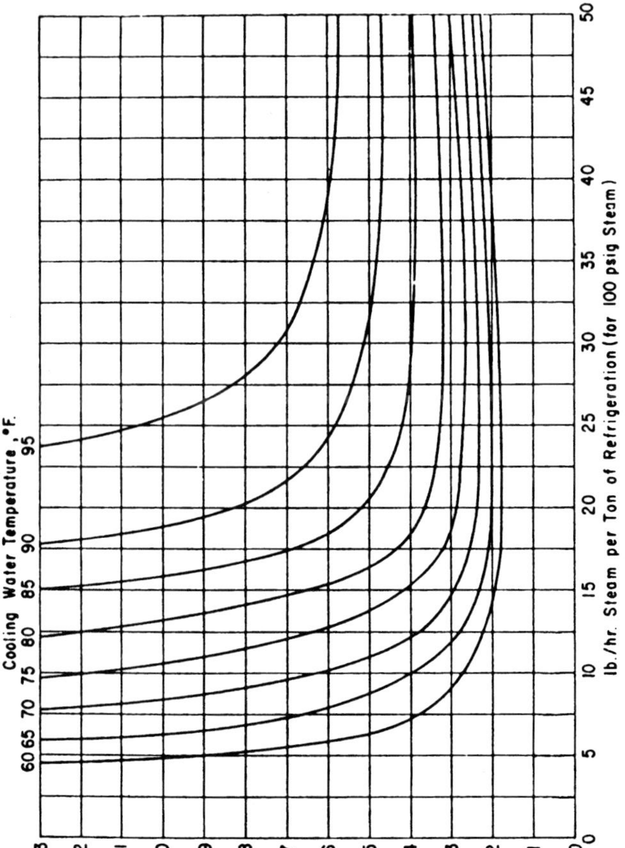

Figure 11-4. lb/hr steam per ton of refrigeration (for 100 psig steam) for 50°F chilled water.

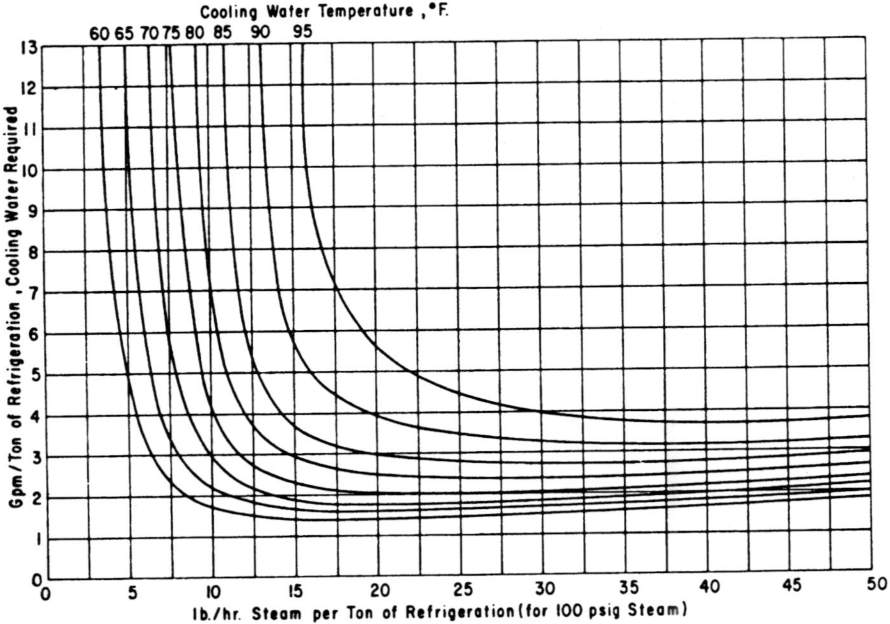

Figure 11-5. lb/hr steam per ton of refrigeration (for 100 psig steam) for 60°F chilled water.

Table 11-3
Steam and Cooling Water Required for Ammonia Absorption Refrigeration Systems[1]

Single-Stage				
Evap. Temp., °F	Steam Sat. Temp., °F. Req. In Generators	Btu Per Min. Req. In Generator Per Ton Refrig. (200 Btu/ Min.)	Steam Rate lb/hr/ton Refrig.	Water Rate Thru Cond. (7.5°F. Temp. Rise), gpm/ton
50	210	325	20.1	3.9
40	225	353	22.0	4.0
30	240	377	23.7	4.1
20	255	405	25.7	4.3
10	270	435	28.0	4.6
0	285	467	30.6	4.9
−10	300	507	33.6	5.4
−20	315	555	37.3	5.9
−30	330	621	42.5	6.6
−40	350	701	48.5	7.7
−50	370	820	57.8	9.5

Table 11-3 *(continued)*

Two-Stage				
Evap. Temp., °F	Steam Sat. Temp., °F. Req. In Generators	Btu Per Min. Req. In Generator Per Ton Refrig.	Steam Rate lb/hr/ton Refrig.	Water Rate Thru Cond. (7.5°F. Temp. Rise), gpm/ton
50	175	595	35.9	4.3
40	180	625	37.8	4.5
30	190	655	40.0	4.6
20	195	690	42.3	4.9
10	205	725	44.7	5.3
0	210	770	47.5	5.7
−10	220	815	50.6	6.3
−20	230	865	54.0	6.9
−30	240	920	58.0	7.8
−40	250	980	62.3	9.0
−50	265	1050	67.5	11.0

(text continued from page 185)

AMMONIA ABSORPTION TYPE UTILITIES REQUIREMENTS

Steam and cooling water requirements for ammonia absorption refrigeration systems are shown in Table 11-3 for single-stage and two-stage units. The tables are based upon cooling water to the condenser of 85°F with 100°F condensing temperature. Water from the condenser is used in the absorbers.

Example. For an evaporator temperature of –10°F, a steam rate (300°F saturated temperature in the generators) of 33.6 lb/hr/ton refrigeration is required. Also, 5.4 gpm cooling water/ton refrigeration, assuming a 7.5°F rise through the condenser, are required in this system.

REFERENCES

1. Ludwig, E. E., *Applied Process Design for Chemical and Petrochemical Plants,* Vol. 3, 3rd ed., Gulf Publishing Co., Houston, Tex., 2000 (in production).
2. Ballou, Lyons and Tacquard, "Mechanical, Refrigeration Systems," *Hydrocarbon Processing,* June 1967.
3. E.I. DuPont de nemours and Company web site (www.dupont.com), "Selection of DuPont Refrigerants," reprinted by permission.
4. Equations were generated using FLEXCURV, Gulf Publishing Co.

12
Tank Blending

This chapter discusses a simple low cost tank blending method. It is referred to as a shrouded blending nozzle system, and it works quite well. The shroud causes the jet nozzle to educt a large quantity of surrounding fluid, improving blending. A simplified diagram of the nozzle portion is shown in Figure 12-1.

Make the nozzle point upward at an angle such that a straight line projected from the nozzle would hit the liquid surface ½ to ⅔ of the way across the tank diameter. The idea is to promote top to bottom turnover. However, tilt the nozzle slightly to the left to promote a slight swirl effect. Aim the nozzle at a point about ⅓ of a radius off-center.

The following rules of thumb apply primarily to the situation where a final tank requires blending after, and perhaps during, transfer from rundown tanks.

1. Base tank size of 10,000 barrels.
2. Use about 25 hp circulating pump.
3. Provide roughly 50–75 feet of head, or a little higher.

4. For a size other than 10,000 barrels, ratio directly for horsepower.
5. Circulation rate in gpm can be calculated from horsepower and head.

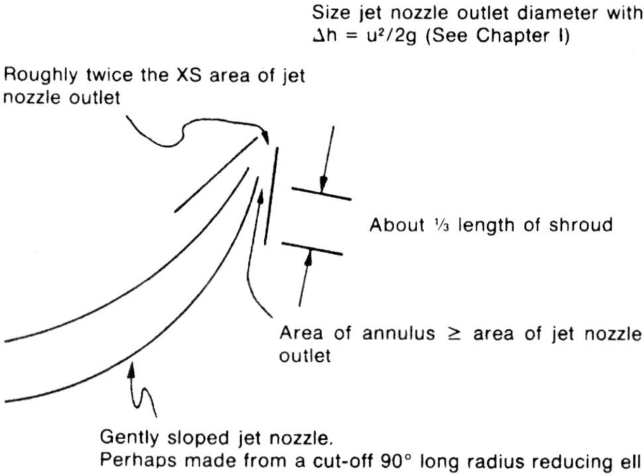

Figure 12-1. Blending can be improved by a shrouded blending nozzle system because the jet nozzle educts a large quantity of surrounding fluid.

Often in refrigerated storage, light hydrocarbons are involved that have a large gravity change with temperature. This condition makes such storage relatively easy to stratify. Be sure not to put in too large a blending pump for refrigerated storage. Not only does it waste blending pump horsepower, but extra heat is added to the fluid that must be removed by the refrigeration system (a double penalty). Operators often tend to leave blending systems running full time.

Provide a means of filling the tank to operating level prior to operating the blending nozzle system. Damage could result from discharging the high velocity jet into an empty tank.

When circulating the blending system and running down into the tank at the same time, it may be possible to direct the rundown stream into the circulating pump suction for additional blending in the pump.

13
Utility Systems

COOLING WATER SYSTEMS

To determine cooling water system flows, use a heat and material balance and a chloride balance (concentration ratio is usually calculated form chloride concentrations).

$$\begin{aligned} D &= C(500)\text{lb / hr}(1)\text{Btu / lb°F}(\Delta T)\text{°F} \\ &= 500\ C\Delta T\ \text{Btu / hr} \end{aligned} \qquad (13\text{-}1)$$

$$E \approx \frac{D\ \text{Btu / hr}}{1000\ \text{Btu / lb}\ (500)\ \text{lb / hr / gpm}} = \frac{C\Delta T}{1000} \qquad (13\text{-}2)$$

$$CR = \frac{Cl_B^-}{Cl_M^-};\ \text{also}\ MCl_M^- = BCl_B^- \qquad (13\text{-}3)$$

So

$$CR = M/B \qquad (13\text{-}4)$$

$$M = E + B\ \text{(overall material balance)} \qquad (13\text{-}5)$$

$$E = M - B = CR(B) - B = B(CR{-}1) \qquad (13\text{-}6)$$

$$B = \frac{E}{CR - 1} \tag{13-7}$$

$$M = E\left(\frac{CR}{CR - 1}\right) \tag{13-8}$$

Use Equation 13-2 to get E, then Equation 13-7 to get B, and finally Equation 13-5 or 13-8 to get M.

Where D = cooling system duty, Btu/hr
C = system circulation rate, gpm
ΔT = cooling system temperature difference (hot return water minus cold supply water), °F
E = cooling system evaporation rate, gpm
CR = cooling system concentration ratio
Cl^- = chloride concentration in the makeup or blowdown
M = cooling system makeup rate, gpm
B = cooling system total blowdown, gpm. This includes both planned blowdown plus cooling system windage (or drift) losses. (Of course, any system leakage counts as part of "planned" blowdown.)

To determine the required amount of planned blowdown, subtract windage losses from B. Use Table 13-1 for windage losses in lieu of manufacturer's or other test data.

Table 13-1
Windage Loss

Type of Cooling Device	Windage Loss as Percentage of System Circulating Rate
Spray pond	3
Atmospheric cooling tower	0.7
Mechanical draft cooling tower	0.2
(Drift eliminators may do better than 0.2)	

When cooling systems are treated, chemicals are sometimes added in shots rather than continuously. Equation 13-9 gives a chemical's half life in a cooling system:

$$T_{1/2} = \frac{S}{B}(.693) \qquad (13\text{-}9)$$

where $T_{1/2}$ = half life, min.
S = system capacity, gal

WATER ALKALINITY

Most water analysis results are rather easily interpreted. However, two simple and useful tests need explanation. These are the P and M alkalinity. The water is titrated with N/30 HCl to the phenolphthalein endpoint at pH 8.3. This is called the P alkalinity. Similar titration to the methyl orange end point at pH 4.3 is called the M alkalinity. They are reported as ppm $CaCO_3$.

This applies to waters having alkalinity caused by one or all of the following:

1. Bicarbonate (HCO_3^-)
2. Carbonate ($CO_3^=$)
3. Hydroxide (OH^-)

In natural waters the alkalinity is usually caused by bicarbonate. Carbonate or hydroxide is rarely encountered in untreated water. The M alkalinity equals the sum of all three forms of alkalinity. The P alkalinity equals ½ the carbonate and all the hydroxide alkalinity. Table 13-2 shows what one can deduce from the P and M alkalinity.

Table 13-2
Comparison of P and M Alkalinities

	OH^-	$CO_3^=$	HCO_3^-
P = O	O	O	M
P = M	M	O	O
P = M/2	O	M	O
P < M/2	O	2P	M–2P
P > M/2	2P–M	2(M–P)	O

BOILER BLOWDOWN CONTROL

The American Boiler Manufacturers' Association (ABMA) has established limits for boiler water composition. The limits are set to help assure good quality steam (for example, limiting silica in the steam to 0.02–0.03 ppm). Blowdown is normally based on the most stringent of these limits shown in Table 13-3.

Table 13-3
ABMA Limits for Boiler Water

Boiler Pressure (psig)	Total Solids (ppm)	Alkalinity (ppm)	Suspended Solids (ppm)	Silica (ppm)
0–300	3,500	700	300	125
301–450	3,000	600	250	90
451–600	2,500	500	150	50
601–750	2,000	400	100	35
751–900	1,500	300	60	20
901–1,000	1,250	250	40	8
1,001–1,500	1,000	200	20	2.5
1,501–2,000	750	150	10	1.0
over 2,000	500	100	5	0.5

EXCESS AIR FOR FIRING

References 4 and 5 give target excess oxygen to shoot for as a guide in heater efficiency improvement. Table 13-4 summarizes the recommended targets.

In an operating plant, the air rate can be adjusted at fixed heat output (constant steam rate for a boiler) until minimum fuel rate is achieved. This is the optimum so long as the warnings below are heeded.

Woodward[4, 5] warns that excess oxygen should not be allowed to get so low that combustibles can get in the flue

Table 13-4
Target Excess Oxygen

	% Excess Oxygen	
Situation	Gas Firing	Oil Firing
Portable analyzer weekly check	5.0	6.0
Permanently mounted oxygen recorder	4.5	5.0
Oxygen recorder with remote manual damper control	4.0	4.5

gas. This can quickly result in lost efficiency and pose a safety hazard. In oil firing, a fuel rich condition can be detected by smoking. In gas firing, however, substantial loss from unburned combustibles can occur before smoking is seen. Woodward believes that when excess oxygen is below 3%, the percentage of combustibles in the flue gas should be monitored.

Reference 6 discusses controlling the excess air at 10% for plants having highly variable fuel supplies by using gas density to control air/fuel ratio. Reference 6 gives the following equations relating theoretical air and density.

$$\text{Fuel gas/air ratio} = 1 : 14.78\ (1 + 0.0921/r_d)\ \text{mass/mass} \tag{13-10}$$

$$\text{Fuel oil/air ratio} = 1 : 0.115\ (X + 3Y)\ \text{mass/mass} \tag{13-11}$$

where r_d =density relative to air; M.W./M.W. air
X = % carbon
Y = % hydrogen

Equation 13-10, the fuel gas equation, does not hold for unsaturated hydrocarbons; however, for small percentages of unsaturates the error is not serious.

High flame temperature and high excess air increase NO_X emissions.

PROCESS EFFICIENCY

With high fuel costs, the search is on for processes with higher thermal efficiency and ways to improve efficiencies of existing processes. One process being discussed lately for its high efficiency is the gas turbine "combined cycle." The gas turbine exhaust heat makes steam in a waste heater boiler. The steam drives turbines, perhaps even a gas turbine steam helper. References 10, 11, and 12 treat this subject and mention alternate equipment hookups, some in conjunction with coal gasification plants.

Reference 2 is a well written report that discusses power plant coal utilization in great detail. It gives a thermal efficiency of 80–83% for modern steam generation plants and 37–38% thermal efficiency for modern power generating plants at base load (about 70%). A modern base load plant designed for about 400 MW and up will run at steam pressures of 2,400 or 3,600 psi and 1,000°F with reheat to 1000°F and regenerative heating of feedwater by steam extracted from the turbine. A thermal efficiency of 40% can be had from such a plant at full load and 38% at high annual load factor. The 3,600-psi case is supercritical and is called a once-through boiler because it has no steam drum. Plants designed for about 100–350 MW run around 1,800 psi and 1,000°F with reheat to 1,000°F. Below 100 MW a typical condition would be about 1,350 psi and 950°F with no reheat. Reference 2 states that below 60%, load factor efficiency falls off rapidly and that the average efficiency for all steam power plants on an annual basis is about 33%.

For any process converting heat energy to mechanical work, the Carnot efficiency is the theoretical maximum. It is calculated as

$$\frac{T_1 - T_2}{T_1} \times 100 \qquad (13\text{-}12)$$

where T_1 = temperature of the heat source, °R
T_2 = temperature of the receiver where heat is rejected, °R

Therefore, the efficiency is raised by increasing the source temperature and decreasing the receiver temperature.

The efficiency for a boiler or heater is improved by lowering its stack temperature. The stack minimum temperature is frequently limited by SO_3 gas dew point. References 7, 8, and 9 discuss this important subject. A stack as hot as 400°F (or perhaps higher) can have problems if the SO_3 concentration is high enough. Reference 14 states that SO_3

condensation will produce a blue-gray haze when viewed against a clear blue sky.

A very useful relationship for determining the maximum available energy in a working fluid is

$$\Delta B = \Delta H - T_0 \Delta S \tag{13-13}$$

where ΔB = maximum available energy in Btu/lb
ΔH = enthalpy difference between the source and receiver, Btu/lb. For a typical condensing steam turbine it would be the difference between the inlet steam and the liquid condensate.
T_o = receiver temperature, °R
ΔS = entropy difference between the source and receiver, Btu/lb °F

To obtain lb/hr-hp, make the following division:

$$\frac{2{,}545}{\Delta B}$$

Equation 13-13 will yield the same result as the Theoretical Steam Rate Tables.[13] Therefore, this is a handy way of getting theoretical steam rates having only a set of steam tables sans Mollier diagram.

ENERGY FROM A GAS EXPANDER

With high energy costs, expanders will be used more than ever. A quick rough estimate of actual expander available energy is

$$\Delta H = C_p T_1 \left[1 - \left(\frac{P_2}{P_1}\right)^{(K-1)/K}\right] 0.5 \qquad (13\text{-}14)$$

where ΔH = actual available energy, Btu/lb
C_p = heat capacity (constant pressure), Btu/lb °F
T_1 = inlet temperature, °R
P_1, P_2 = inlet, outlet pressures, psia
K = C_p/C_v

To get lb/hr-hp, divide as follows:

$$\frac{2{,}545}{\Delta H}$$

A rough outlet temperature can be estimated by

$$T_2 = T_1 \left(\frac{P_2}{P_1}\right)^{(K-1)/K} + \left(\frac{\Delta H}{C_p}\right) \qquad (13\text{-}15)$$

For large expanders, Equation 13-14 may be conservative. A full rating using vendor data is required for accurate results. Equation 13-14 can be used to see if a more accurate rating is worthwhile.

For comparison, the outlet temperature for gas at critical flow across an orifice is given by

$$T_2 = T_1\left(\frac{P_2}{P_1}\right)^{(K-1)/K} = T_1\left(\frac{2}{K+1}\right) \qquad (13\text{-}16)$$

The proposed expander may cool the working fluid below the dew point. Be sure to check for this.

INVERTED BUCKET STEAM TRAPS

It is easy to improperly design a steam trap. The design must work for two circumstances and often a designer will check only one of these. The circumstance often overlooked is as follows: On startup or upset, the steam control valve can open wide so that the steam chest pressure (assume for this discussion that we are speaking of a reboiler) rises to full steam line pressure. At a time like this, the steam trap downstream pressure can be atmospheric due to process variations or the operators opening the trap discharge to atmosphere in an attempt to get it working.

If the trap orifice has been designed too large, the trap valve cannot open to discharge condensate, creating or amplifying serious plant problems. For any steam trap, for a given trap pressure differential, there is a maximum orifice size above which the bucket can't exert sufficient opening power for the trap to operate. So when designing a trap, check manufacturers' data to stay within the maximum sized orifice for full steam line pressure to atmospheric. If a larger orifice is required by the alternate circumstance discussed below, a larger trap body size must be specified whose bucket can service the larger orifice.

The required orifice continuous flow capacity is determined at steam chest pressure to condensate system pressure at a flow 6 to 8 times design. If designed for normal flow, the trap would have to be open 100% of the time. Then, as stated above, a body size is selected that can contain the required orifice (not be above the stated maximum for that body size in the manufacturer's sizing tables) at the condition of full steam line pressure to atmospheric. The various vendor catalogs provide sizing charts and tables.

MIXING HIGH- AND LOW-PRESSURE CONDENSATE

Figure 13-1 shows high-pressure condensate (small line) being added to low-pressure condensate without the usual troublesome "hammer." The high-pressure condensate has a chance to cool before emerging into the low-pressure condensate line.

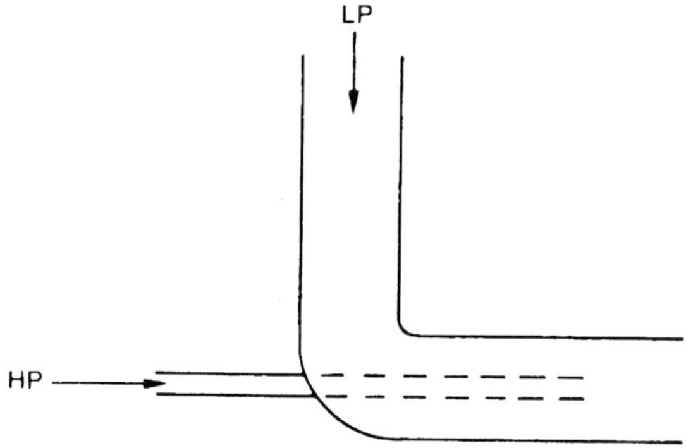

Figure 13-1. This method is recommended for mixing high- and low-pressure condensates.

HEATING WATER WITH STEAM

The usual steam-in-shell versus water-in-tubes heat exchanger is best designed as follows:

1. Vertical unit.
2. Control valve in outlet condensate.
3. Excess area to allow control of heat balance by varying the level of condensate at the bottom of the tubes.

The above unit is economical, avoids water hammer, has excellent control, and is easy to start up. Contrary to popular belief, heating water with steam is not a highly straightforward design problem. This unit eliminates problems with other designs.

REFERENCES

1. *Betz Handbook of Industrial Water Conditioning,* 6th ed., Betz Laboratories Inc., Trevose, Pa., 1962.
2. Locklin, D. W., H. R. Hazard, S. G. Bloom and H. Nack, *Power Plant Utilization of Coal,* A Battelle Energy Program Report, Battelle Memorial Institute, Columbus, Ohio, Sept. 1974.
3. *NGPSA Engineering Data Book,* Natural Gas Processors Suppliers Association, 9th ed., 1972.
4. Woodward, A. M., "Control Flue Gas to Improve Heater Efficiency," *Hydrocarbon Processing,* May 1975.
5. Woodward, A. M., "Reduce Process Heater Fuel," *Hydrocarbon Processing,* July 1974.
6. Ferguson, B. C., "Monitor Boiler Fuel Density to Control Air/Fuel Ratio," *Hydrocarbon Processing,* Feb. 1974.
7. Verhoff, F. H. and J. T. Banchero, "Predicting Dew Points of Flue Gases," *Chemical Engineering Progress,* Vol. 70, No. 8, Aug. 1974.
8. Martin, R. R., F. S. Manning, and E. D. Reed, "Watch for Elevated Dew Points in SO_3-Bearing Stack Gases," *Hydrocarbon Processing,* June 1974.

9. "Fuel Additives Control Preburner and Fireside Combustion Problems," Betz Bulletin 713, Betz Laboratories, Inc., 1974.
10. Moore, R. and C. Branan, "Status of Burnham Coal Gasification Project," *Proceedings, 54th Annual Convention, Gas Processors Association,* Houston, Tex., March 10–12, 1975.
11. Zanyk, J. P., "Power Plant Provides 86% Efficiency," *Oil and Gas Journal,* May 27, 1974.
12. Ahner, D. J., T. S. May, and R. C. Sheldon, "Low Btu Gasification Combined-Cycle Power Generation," presented at Joint Power Generation Conference, Miami Beach, September 15–19, 1974.
13. Keenan, J. H., and F. G. Keyes, "Theoretical Steam Rate Tables," *Trans. A.S.M.E.,* 1938.
14. Reed, R. D., "Recover Energy from Furnace Stacks," *Hydrocarbon Processing,* Jan. 1976.

14
Pressure Drop in a Catalyst Bed

Equation 14-1 is a method developed by Girdler Chemical, Inc. for calculating the pressure drop in a catalyst bed. The author has verified it in actual plant operation.

$$\Delta P = 6.58 \times 10^{-10} \times R \times G^{1.9} \times Q \times L/(D_p^{1.1} \times \rho) \qquad (14\text{-}1)$$

where ΔP = pressure drop, psi

R = roughness factor. (Smooth surfaces such as glass = 1.0, very rough surfaces = 2.3, Girdler G-56, G-3 = 1.5)

G = mass flow, $lb/hr/ft^2$ of open cross section

L = bed depth, ft

ρ = gas density, lbs/ft^3 based on the average of inlet and outlet temperatures, pressures, and volumes

D_p = 6 V/A

V = actual volume of catalyst pellet or ring, $in.^3$ (For rings, does not include volume of hole.)

Table 14-1
Particle Diameter

D_o	L	D_i	D_p	$D_p^{1.1}$
¾	¾	5/16	0.5081	0.475
⅝	⅝	¼	0.4326	0.398
⅝	⅝	5/16	0.3750	0.339
⅝	⅜	¼	0.3750	0.34
⅝	⅜	5/16	0.3309	0.296
⅝	¼	¼	0.3214	0.286
⅝	¼	5/16	0.2884	0.255

A = total surface area, in.^{2}
$Q = (1 - E_p) / E_p^3$
$E_p = E_s + (1 - E_s)(D_i / D_o)^2$
D_i = inside diameter of pellet, in.
D_o = outside diameter of pellet, in.
$E_s = 0.373 + 0.308\,(D_o / D_t)$
D_t = inside tube or reactor diameter, in.

Limitation:

1. $D_pG/12\mu > 200$, μ = fluid viscosity, lb/ft hr
2. ΔP should not be greater than 10–15% of inlet pressure. The particle diameter, D_p, can be determined from Table 14-1.

REFERENCE

1. Information from Girdler Chemical, Inc., Louisville, Kentucky, 40201.

15
Weighted Spray

A friend developed a unique spray that holds very constant velocity and consistent spray pattern regardless of flow, thereby overcoming turndown problems in other spray nozzles. This is very useful in quench or desuperheating service. Figure 15-1 illustrates this unique spray.

Using the pressure drop across the area producing spray in lbs/in^2 and the lid contact surface in in^2, calculate the weight in pounds. The pressure drop will be the pressure provided to the tee minus the vessel pressure.

The spray velocity can be calculated from $\Delta h = u^2 / 2g$, the units of which were previously defined. The velocity in ft/sec divided into the horizontal spray distance in feet gives the horizontal time component. The spray fall over that length of time is

$$S = \tfrac{1}{2} g T^2 \qquad (15\text{-}1)$$

where S = fall distance, ft
T = time, sec
g = 32.2 ft/sec^2

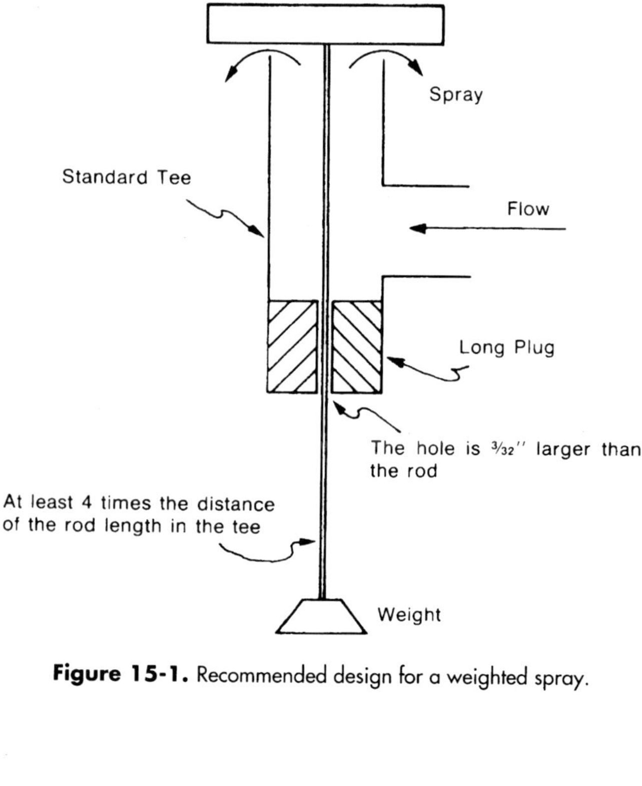

Figure 15-1. Recommended design for a weighted spray.

16
Properties and Conversion Factors

INTRODUCTION

There are many sources of common properties of hydrocarbons and well-known chemicals. The more difficult to find properties are therefore provided in this pocket guide.

APPROXIMATE PHYSICAL PROPERTIES

Gravity of a boiling hydrocarbon liquid mixture at all pressures:

4 lb/gal or 30 lb/ft^3

Liquid volume contribution of dissolved gases in hydrocarbon mixtures:

Component	MW	gals/lb mol
O_2	32	3.37
CO	28	4.19
CO_2	44	6.38
SO_2	64	5.50
H_2S	34	5.17
N_2	28	4.16
H_2	2	3.38

Liquid volume contribution of light hydrocarbons in hydrocarbon mixtures:

Component	MW	gals/lb mol
C_1	16	6.4
$C_2^=$	30	10.12
C_2	28	9.64
$C_3^=$	44	10.42
C_3	42	9.67
iC_4	58	12.38
nC_4	58	11.93
iC_5	72	13.85
nC_5	72	13.71
iC_6	86	15.50
C_6	86	15.57
iC_7	100	17.2
C_7	100	17.46
C_8	114	19.39

Higher heating values of various fuels:

Solid Fuels	HHV, Btu/lb
Low volatile bituminous coal and high volatile anthracite coal	11,000–14,000
Subbituminous coal	9,000–11,000
Lignite	7,000–8,000
Coke and coke breeze	12,000–13,000
Wood	8,000–9,000
Bagasse (dry)	7,000–9,000

Gaseous Fuels	HHV, Btu/scf
Coke oven gas	500–700
Blast furnace gas	90–120
Natural gas	900–1,300
Refinery off gas	1,100–2,000

Liquid Fuels	HHV, Btu/gal
Heavy crude (14°API)	150,000
Light crude (30°API)	140,000
Fuel Oils No. 1	137,400
No. 2	139,600
No. 3	141,800
No. 4	145,100
No. 5	148,800
No. 6	152,400
15°API	178,710
Kerosine (45° API)	159,000
Gasoline (60° API)	150,000
Naphtha (65° API)	115,000
Butane	103,300
Propane	91,600
Ethane	66,000

Assumed gravities used commercially in determining rail tariffs:

Propane or butane	4.7 lb/gal
Natural gasoline	6.6 lb/gal

Volume of a "perfect gas":

76 ft3/lb mol at 76°F and 76 psia

This volume is directly proportional to the absolute temperature = °F plus 460. It is inversely proportional to the absolute pressure.

Molal heat of vaporization at the normal boiling point (1 atm):

8,000 cal/gm mol or 14,400 Btu/lb mol

Compressibility factor (Z) for mixtures when using pseudo-critical mixture constants to determine it.

Add 8°C to the critical temperature and 8 atmospheres to the critical pressure of hydrogen and helium.

The latent heat of water is about 5 times that for an organic liquid.

Heat capacities, Btu/lb°F	
Water	1.0
Steam/hydrocarbon vapors	0.5
Air	0.25
Solids	0.25

For structural strength:

Weight of concrete = Weight of equipment supported
Density of concrete = 2 × Density water
Density of steel = 8 × Density water

For desert landscaping (home or office)—contractors use:
yards (yds^3) rock = ft^2 covered/100

GAS SOLUBILITY

Henry's Law states that:

$$p = Hx$$

where p = partial pressure of the solute in the gas phase, atm
x = mole fraction of the solute in the liquid phase
H = the Henry's Law constant

Example: Calculate the solubility of oxygen in water at ambient conditions. Assume air-water contact at 20°C (68°F) and sea level (1 atmosphere).

p = .21 atm (oxygen is 21 mol% in air)
$H = 4.01 \times 10^4$ from Perry's
$x = p/H = .21/40{,}100 = 5.24 \times 10^6$ or 5.2 ppm molar

By weight oxygen is $5.2 \times 32/18 = 9.2$ ppm

DENSITY OF IRREGULAR SOLIDS

I am sure all readers know the story of Archimedes and the king's crown, however, review is instructive. King Hiero heard the rumor that the goldsmith had substituted an equal weight of silver for part of the gold supplied for the crown. He asked Archimedes to find out if the crown was pure gold.

Archimedes developed his solution while getting into a full bathtub. As the water rose he jumped from the bath shouting, "Eureka!" ("I have found it!")

In the first century B.C. the Roman architect, Vitruvius, told a story about a method Archimedes apparently used to investigate the fraud. The method involved putting a pure gold nugget, equal in weight to the crown, into a bowl of water. After the bowl was filled to the brim, the nugget was removed and the crown put in its place. The crown was a lighter alloy, and therefore its greater volume caused the bowl to overflow.

The Drexel University Website (www.mcs.drexel.edu) discusses the relative inaccuracy of the proposed Vitru-

vius method and gives a more plausible method befitting Archimedes' great intelligence and using his Law of Buoyancy.

The better method involves counterbalancing the nugget and the crown suspended from a beam and submerging both nugget and crown in water. The crown would displace more water than the nugget and thus rise while the nugget sank.

At Sasol in South Africa, a similar technique is used to determine the density of coal fed to the gasifiers. The coal is first weighed in air and then weighed submerged in water. The apparent loss of mass in water equals the mass of water displaced. From the known density of water, the volume of water is determined. Because a submerged object displaces its volume of water, the volume of coal is equal to the volume of displaced water. Finally, dividing the weight in air by the volume gives the density of the coal. This is used as a parameter in setting gasifier operation.

For good stories about Archimedes, see the aforementioned Website or read *A History of* π by Petr Beckman, St. Martin's Press, New York.

APPROXIMATE CONVERSION FACTORS

Conversion factors can be easily remembered if altered slightly, but not significantly enough to affect shortcut calculations. Here are some examples:

Factor		Actual	Remember As
R (gas constant)*	Btu/lb mol°R	1.987	2
	ft3psi/lb mol°R	10.74	see footnote
ft lb/Btu		778	777
°F/°C or vice versa		Two formulas are commonly used	C/5 = (F–32)/9
Hydrocarbon mol wts		Usually to 2 decimal places	Carbons × 12 plus Hydrogens × 1
Quarts/Liter		1.05671	1
ft^3/lb mol perfect gas at 76 psia and 76°F		75.56	76
Km/mile		1.609347	1.6
mile/Km		0.6213699	0.6
lb/Kgm		2.204622	2.2
Value of $Q/D^{2.5}$ above which a horizontal pipe is flowing full (Q=gpm, D=diameter, in.)		10.2	10

* Cal/gm mol°K = Btu/lb mol°R

To go to other values of R such as ft^3psi/lb mol°R, just use memorized factors:

2Btu/lb mol°R × 777 ft-lb/Btu × 1 ft^3 (lb/ft^2)/ft-lb × psi/144 (lb/ft^2) = 10.79 ft^3 psi/lb mol°R

REFERENCES

1. Branan, C., *Rules of Thumb for Chemical Engineers,* 2nd ed., Gulf Publishing Co., Houston, TX, 1998.
2. Perry, R. H. and D. W. Green, *Perry's Chemical Engineering Handbook,* 7th ed., McGraw-Hill, 1997.